Amino Acid Analysis by Gas Chromatography

Volume III

Editors

Robert W. Zumwalt

Kenneth C. T. Kuo

Charles W. Gehrke
University of Missouri-Columbia
Columbia, Missouri

CRC Press, Inc.
Boca Raton, Florida

Library of Congress Cataloging-in-Publication Data

Amino acid analysis by gas chromatography.

  Bibliography: p.
  Includes index.
  1. Amino acids. 2. Gas chromatography. I. Zumwalt,
Robert W. II. Kuo, Kenneth C. T., 1936-
III. Gehrke, Charles W.
QD431.A656 1987 Vol 3 547.7'5   86-9649
ISBN 0-8493-4328-3 (Set)
ISBN 0-8493-4329-1 (v. 1)
ISBN 0-8493-4330-5 (v. 2)
ISBN 0-8493-4331-3 (v. 3)

71870

Direct all inquiries to CRC Press, Inc., 2000 Corporate Blvd., N.W., Boca Raton, Florida, 33431.

© 1987 by CRC Press, Inc.

International Standard Book Number 0-8493-4328-3 (Set)
International Standard Book Number 0-8493-4329-1 (Volume I)
International Standard Book Number 0-8493-4330-5 (Volume II)
International Standard Book Number 0-8493-4331-3 (Volume III)

Library of Congress Card Number 86-9649
Printed in the United States

In the background the remaining columns of Academic Hall, first building of the University of Missouri, and Jesse Hall, administrative building for the University of Missouri-Columbia, presenting:

Dr. Charles W. Gehrke (center), Professor of Biochemistry and Director of the University of Missouri Interdisciplinary Chromatography Mass-Spectrometry Facility and Centennial President (1984) of the International Association of Official Analytical Chemists, Co-Principal Investigator on lunar analysis, Apollos 11 to 17, in search for life molecules.

Dr. Robert W. Zumwalt (right), Research Associate in Biochemistry and Co-Investigator on lunar analysis team of Apollos 11 to 17.

Mr. Kenneth C. T. Kuo (left), Senior Chromatographer and Research Chemist, and Co-Investigator on lunar analysis team of Apollos 11 to 17.

THE EDITORS

Robert Wayne Zumwalt was born in 1944 in southwest Missouri near the town of Bolivar. After attending Polk and Bolivar public schools, he entered Southwest Missouri State University and received the B.S. degree in chemistry in May 1966.

In June of that year he began graduate studies at the University of Missouri in Columbia as a research assistant in the laboratory of Prof. Charles W. Gehrke. He received the M.S. degree in 1968, after performing thesis research on the separation characteristics of polyester liquid phases, the effects of heat treatment of support materials on the chromatography of the amino acid N-TFA n-butyl esters, and the synthesis and chromatography of the trimethylsilyl derivatives of the amino acids.

Dr. Zumwalt was awarded a National Science Foundation traineeship in 1968, and his doctoral thesis research concerned both the extension of the N-TFA n-butyl ester method for analysis of complex biological materials and the development and application of high sensitivity methods for examination of the returned lunar samples for indigenous amino acids. He participated in the analysis of lunar samples returned by Apollo lunar missions for amino acids and extractable organic compounds at the NASA Ames Research Center, California, and the Laboratory for Chemical Evolution at the University of Maryland. He was co-inventor of a solvent venting system for gas chromatography (GC), and conducted the first derivatization and chromatographic studies using *bis*-(trimethylsilyl)trifluoroacetamide as a silylating agent for amino acids.

He, along with Dr. Gehrke and Mr. Kuo, reported the dual-column chromatographic system which allowed the quantitative analysis of the protein amino acids as the N-TFA n-butyl esters, and completed his thesis research with the development of ion-exchange purification procedures which allowed amino acid analysis of physiological fluids and other complex materials.

After receiving the Ph.D in 1972, Dr. Zumwalt's research efforts turned to the development of analytical methods for detecting and measuring biological markers of neoplastic disease. These studies included development of GC and HPLC techniques for measurement of methylated bases and nucleosides in patients physiological fluids. Dr. Zumwalt was associated with the College of Veterinary Medicine, University of Missouri, for 4 years, with the U.S. Fish and Wild Life Service's Columbia National Fisheries Research Laboratory for 1 year, then returned to the University of Missouri Department of Biochemistry in 1981. Dr. Zumwalt lectures in a graduate level course in Analytical Biochemistry-Chromatography at the University of Missouri-Columbia, and is an author of more than 50 scientific publications in the field of the development and application of quantitative chromatographic methods in biochemical research.

Kenneth Ching-Tien Kuo, was born in 1936 in China. He studied at Chun-Yen Institute of Science and Engineering, Taiwan, receiving a B.S. degree in Chemical Engineering in 1960. After fulfilling a military service obligation, he enrolled at the University of Houston. In 1963 he joined the Chevron Chemical Co. in Richmond, California, developing pesticide residue analytical methods and studying pesticide metabolism. Recognizing the power of gas-liquid chromatography (GLC) and the need of high resolution, sensitivity, and speed in the analysis of amino acids, he applied and was accepted as a member of the research team under Prof. Charles Gehrke at the University of Missouri-Columbia in 1968. He developed mixed phase columns for histidine, arginine, and cystine, which allow the dual column complete quantitation of protein amino acids in 30 min by GC. He along with Drs. Gehrke, Stalling, and Zumwalt, invented the Solvent-Vent Chromatographic System (U.S. Patent No. 3,881,892) which eliminates the sample solvent effect in GC analysis. This solvent-venting device was used in the search for amino acids in the returned Apollo lunar samples over the period from 1969 to 1974, thus providing a sensitivity factor of 100 greater than classical ion-exchange analysis at that time. He received his M.S. degree in analytical biochemistry under Prof. Gehrke in 1970.

During the last 17 years he and Dr. Gehrke have dedicated their research efforts to the developments of quantitative high resolution chromatographic methods for biochemical and biomedical research. He participated in the NASA Apollo Returned Lunar Sample consortium of scientists searching for evidence of chemical evolution in the lunar samples from Apollo missions 11 through 17 (1969 to 1974). He has studied biomarkers for cancer, and developed quantitative high resolution chromatographic methods for polyamines, protein-bound neutral sugars, β-aminoisobutyric acid, and β-alanine; and modified ribonucleosides in human urine and serum. In the last 5 years his major efforts have been directed to the developing of a package of methods for the complete quantitative composition analysis of DNA, mRNA, and tRNA by high resolution HPLC. Through these methods, more than 70 major and modified ribonucleosides, 15 deoxynucleosides, and 9 mRNA cap structures can be identified and measured in nucleic acids or body fluids. He was an invited scientist by the Chinese Academy of Sciences in 1982 and lectured throughout China on the chromatography of nucleosides. He has contributed to over 50 scientific publications in analytical chemistry and biochemistry.

Charles William Gehrke was born in 1917 in New York City. He studied at the Ohio State University receiving a B.A. in 1939, a B.S. degree in education in 1941, and a M.S. degree in 1941. From 1941 to 1945 he was professor and chairman of the Department of Chemistry at Missouri Valley College, Marshall, Missouri, teaching chemistry and physics to selected Navy midshipmen from the destroyers, battleships, and aircraft carriers of World War II in the South Pacific. These young men then went back to the war theater as Deck and Flight Officers. In 1946, he returned to the Ohio State University as an instructor in agricultural biochemistry and received his Ph.D. degree in 1947. In 1949 he joined the College of Agriculture at the University of Missouri-Columbia, where at present he is professor of biochemistry and manager of the Experiment Station Chemical Laboratories and Director of the University Interdisciplinary Chromatography Mass-Spectrometry Facility. His duties also include those of State Chemist for the Missouri Fertilizer and Limestone Control Laws.

Prof. Gehrke is the author of over 250 scientific publications in analytical and biochemistry. His research interests include the development of quantitative, high-resolution GC and LC methods for amino acids, purines, pyrimidines, major and modified nucleosides in RNA, DNA, and methylated "CAP" structures in mRNA; fatty acids, and biological markers in the detection of cancer; characterization and interaction of proteins, chromatography of biologically important molecules, and automation of analytical methods for nitrogen, phosphorus, and potassium in fertilizers. Automated spectrophotometric methods have been developed for lysine, methionine, and cystine.

Prof. Gehrke has been an invited scientist to lecture on GLC of amino acids in Japan, China, and at many universities and institutes in the U.S. and Europe. He participated as a co-principal investigator in the analysis of lunar samples returned by Apollo flights 11, 12, 14, 15, 16, and 17 for amino acids and extractable organic compounds with Prof. Cyril Ponnamperuma, University of Maryland, and with a consortium of scientists at the NASA Ames Research Center, California. In 1971, he received the annual Association of Official Analytical Chemists' (AOAC) Harvey W. Wiley Award in Analytical Chemistry and was the recipient of the Senior Faculty Member Award, UMC College of Agriculture, in 1973. In August 1974, he was invited to the Soviet Academy of Sciences to make the summary presentation on organic substances in lunar fines to the Oparin International Symposium on the "Origin of Life." In 1975, he was selected as a member of the American Chemical

Society Charter Review Board for Chemical Abstracts. As an invited teacher under the sponsorship of five Central American governments, he taught chromatographic analysis of amino acids at the Central American Research Institute for Industry in Guatemala, 1975.

He was elected to Who's Who in Missouri Education, 1975, and recipient of the Faculty-Alumni Gold Medal Award and was the recipient of the prestigious Kenneth A. Spencer Award from the Kansas City Section of the American Chemical Society for meritorious achievement in agricultural and food chemistry, 1979 to 1980. He received the Tswett "Chromatography Memorial Medal" from the Scientific Council on Chromatography, Academy of Sciences of the USSR, Moscow, 1978, and Sigma Xi Senior Research Award by the University of Missouri-Columbia Chapter, 1980. He has been an invited speaker on "Modified Nucleosides and Cancer" in Freiburg, West Germany, 1982, and gave presentations as an invited scientist throughout Japan, mainland China, Taiwan, Philippines, and Hong Kong, 1982. He was selected to the Board of Directors and Editorial Board of the AOAC from 1979 to 1980, President-Elect of the AOAC international organization, 1982 to 1983; and was honored by the election as the centennial President, 1983 to 1984. He developed "Libraries of Instruments" interdisciplinary research programs on strengthening research in American universities. Dr. Gehrke is founder and chairman of the Board of Directors, Analytical Biochemistry Laboratories, Inc., 1968 to present, a private corporation of 150 scientists, engineers, biologists, and chemists specializing in chromatographic instrumentation, and addressing problems worldwide to the environment.

Over 60 masters and doctoral students have received their advanced degrees in analytical biochemistry under the direction of Prof. Gehrke. In addition to his extensive contributions to amino acid analysis by GC, Dr. Gehrke and colleagues have pioneered in the development of sensitive, high-resolution, quantitative HPLC methods for over 80 major and modified nucleosides in RNA, DNA, mRNA, then applied their methods in collaborative research with scientists in molecular biology across the world. Prof. Ernest Borek at the International Symposium on Cancer Markers, Freiburg, West Germany, in 1982, stated that Prof. Gehrke's chromatographic methods are being used successfully by more than half of the scientists in attendance at these meetings.

In 1986, Dr. Gehrke was the recipient of the American Chemical Society's Midwest Award for outstanding research in analytical biochemistry. The Smithsonian Institution has requested the gas chromatograph which Gehrke and colleagues modified and used to analyze the Apollo lunar samples for display in the National Aeronautics Air and Space Museum.

PREFACE

The central role of proteins (French *protéine,* "primary substance [of the body]" from Greek, *prōtos,* first) and their building blocks, the amino acids, in biology has evoked intense and continued interest in protein and amino acid chemistry by scientists representing a wide spectrum of disciplines. The array of substances subjected to examination for their amino acid content is therefore extraordinarily broad, ranging from meteorites and lunar samples to newly synthesized or isolated peptides and proteins. Over the past 4 decades, chromatographic techniques have emerged as the dominant means of amino acid determination, and milestones of that development are apparent. Partition chromatography of the *N*-acyl amino acids was performed by Nobel Laureates A. J. P. Martin and R. L. M. Synge (1941), working in the laboratories of the Wool Industries Research Association in Leeds, England. They addressed the problem of analyzing the amino acids in protein hydrolysates and laid the theoretical foundation on which liquid-liquid chromatography is based. Nobel Laureates Stanford Moore and William Stein, along with D. H. Spackman at the Rockefeller University, New York, pioneered the elegant automated ion-exchange amino acid analysis (1958) which has had a profound impact on amino acid and protein research. In 1952, Nobel Laureates A. T. James and A. J. P. Martin described the fundamental parameters of gas-liquid chromatography (GLC), including a theory of its operation in terms of the theoretical plate concept, and laid the foundation for further development of the technique.

The earliest gas chromatographic (GC) method for analysis of amino acids was described by Hunter et al. in 1956, and involved separation of the aldehydes which resulted after decarboxylation and deamination with ninhydrin. In 1958, Bayer reported the GC separation of the *N*-trifluoroacetyl (*N*-TFA) methyl esters, and *N*-acylated amino acid esters have subsequently emerged as by far the most widely used class of derivatives. In the 1960s, Charles W. Gehrke and his doctoral students William Lamkin and David Stalling at the University of Missouri-Columbia, laid the foundations that resulted in the synthesis of the reference standard derivatives and in establishing the organic reaction and chromatographic separation conditions for the first quantitative amino acid analysis by GLC of the 20 *N*-TFA *n*-butyl esters. Their research on GLC methods for amino acids led to intensive research in more than 100 laboratories across the world directed to studies on the merit of different derivatives, chromatographic columns, detectors, and applications to research in medicine, agriculture, and the environment.

Numerous reports of GLC techniques for amino acid determination began to emerge in the late 1960s, mainly spurred by the offer of improved resolution and speed of analysis as compared to the ion-exchange techniques of the day. Advances in GC detectors, column, materials, and quantitative derivatization methods during the 1960s and 1970s encouraged further research; and continued improvements in all phases of GC instrumentation and column technology into the 1980s have enhanced the capabilities of GC for amino acid analysis. Development of the GC methodology was followed by interfaced GC/mass spectrometric (MS) analysis and characterization of unknowns and analysis of amino acids enantiomers, and then by the more recent reversed-phase liquid chromatography approaches. This story on accomplishments continues with the excellent contributions of the pioneering scientists in the 20 chapters of this three volume treatise. Analytical and chromatographic strategies for separating, identifying, and quantitating amino acids in the array of matrices has been varied, dictated by both the methodology available and the demands presented by the specific analytical problem whether in a research setting or for compositional information. Improved ion-exchange, gas-liquid, and reversed-phase chromatographic techniques continue to evolve to meet the ever-increasing demands for better resolution, sensitivity, speed, and versatility.

The methodology of choice depends on the analytical requirements at hand. For a protein chemist involved in structural analysis of a particular protein, the analytical demands are not the same as for a nutritional chemist involved in determining the nutritional quality of foods and feeds, the clinical chemist engaged in determining amino acids in physiological fluids to aid in diagnosis of disease, or the biogeochemist interested in the extent of racemization of amino acids in fossils.

The complexity and diversity of the sample matrices that are encountered requires a methodology providing high resolution, selectivity, and a wide linear response range of 10^6. GC/FID (flame ionization detection) is the method of choice and in these situations GC will provide more reliable data. The inherent strengths of GC methods (resolution, sensitivity, versatility, cost) to a wide range of amino acid analytical problems and applications are shown in these volumes.

In biomedical research, the problem is a general one, the need of new techniques and their application to solve old problems, and to probe new ideas of approach to solve intractable new problems. Whatever the disease or biochemical research objective, a research tool is required that will provide a reliable measurement of the molecules under study. GLC of amino acids in all of its ramifications provides the research scientist with powerful new tools and approaches to help in obtaining the needed answers to advance science.

The chromatographies and separation science, are a major "bridge" or "common denominator" as analytical methods in biological sciences research. The importance of research and new methods of measurement to the advancement of our society and the developing world depends upon expanding and new knowledge from every source for continued growth. Problems in nutrition, pollution, drugs, environment, and biotechnology are now being solved by chromatography and interfaced MS in weeks and months, where formerly years of study were involved. The genius of Mikhail Tswett, the father of chromatography, of the early 1900s has had a profound impact to this point in history. To illustrate the significance of Tswett's work, 56 world-leading chromatographers payed tribute to his accomplishments by contributing chapters to *75 Years of Chromatography — A Historical Dialogue*, which was published in 1979 to commemorate the 75th anniversary of Tswett's invention of chromatography. That volume, edited by L. S. Ettre and A. Zlatkis, provides a unique historical perspective as it relates developments and applications of chromatography by scientists from disciplines that range from petroleum chemistry to medicine. His accomplishments promises to open even wider doors as Chemistry and Biology are brought more closely together to more effectively serve mankind.

In 1954, Prof. William Albrecht, Chairman of Agronomy and Soils, at the University of Missouri, expressed the great need and asked me (CWG) to develop a more reliable method for the quantitative measurement of amino acids than the bacterial turbidity method used at that time. This was the challenge and start of my work on new methods by GC for amino acids. Our accomplishments have been most rewarding.

Our goals, at Missouri, in the Experiment Station Chemical Laboratories, have been the development of automated analytical, and chromatographic methods as "research tools" useful for advancing investigation in biochemistry, agriculture, space sciences, and medicine.

Charles W. Gehrke
Kenneth C. Kuo
Robert W. Zumwalt
Columbia, Missouri

A DEDICATION AND THANKS

I humbly dedicate this three-volume treatise on amino acid analysis by gas-liquid chromatography to my beloved son, Dr. Charles W. Gehrke, Jr., a Navy aerospace surgeon whose life was so early and tragically taken in the line of duty on March 1, 1982, at Pensacola, Florida, flight station. Charles was also my graduate student, colleague, closest friend, and an accomplished analytical biochemist. He truly understood the deep meaning and relationships of chemistry and medicine and enjoyed life to its fullest.

I further dedicate these three volumes to my 60 master of science and doctoral students, postdoctorals, colleagues, and visiting scientists, many of whom contributed significantly in the research to a number of chapters of the three volumes. Their efforts have been exemplary and their contributions to analytical biochemistry were meritorious.

Lastly, but not least, my special appreciation to my wife, Virginia, who so graciously supported and encouraged me during these past 3 years during the development and completion of this work.

To my son, Dr. Jon C. Gehrke, M.D. and daughter, Susan G. Gehrke, J.D. for their love and special understanding.

Charles W. Gehrke

ACKNOWLEDGMENTS

Our sincere thanks are extended to the many accomplished scientists from around the world who have graciously and diligently presented their research findings in their contributed chapters. Their efforts have allowed these volumes to present an international, and a comprehensive perspective of the field of gas chromatographic amino acid analysis. In this fast-developing science and technology, their research findings have, and will, play important roles in the advancement of science in many disciplines.

We thank Nancy Rice, Jennifer Welch, Lori Sampsel, and Wylonda Walters Jones for their preparation of finished manuscripts which constitute several chapters of these volumes, and our thanks also to the editors of CRC Press, Inc., for their help in the publication of these volumes.

GENERAL INTRODUCTION

As a result of investigations to answer a wide variety of scientific questions, development of chromatographic techniques for determination of amino acids has advanced continuously since A. J. P. Martin and R. L. M. Synge described partition column chromatography and its application to amino acids in 1941. This wide-ranging interest has had two effects which are apparent in the literature on this subject. First, nearly every form of chromatography has been brought to bear, at one time or another, on problems dealing with the measurement of amino acids. Second, this broad interest has resulted in chromatographic methods for determining amino acids in a broad spectrum of sample types from artificial sweeteners to meteorites.

We undertook this three-volume work to provide the scientific community with information on the development of new research tools and an opportunity to explore the application of packed column and capillary column gas chromatography (GC) to amino acid analysis in a number of different research areas. Packed columns possess certain advantages which include ease of preparation and operation, larger sample capacity, and reduced cost. Capillary columns on the other hand offer increased resolution, selectivity and sensitivity, while fused-silica bonded-phase columns are especially inert to reactive sample components.

VOLUME I — INTRODUCTION

Chapter 1

The author-editors begin with a chapter directed to the critical first step in obtaining accurate measurement of the amino acid content of proteinaceous substances; sample preparation and protein hydrolysis. The main sources of variance between the amino acid content of the HCl hydrolysate and the protein from which it was derived are discussed, and results are presented from studies which compared high-temperature, short-time hydrolysis (145°—4 hr) with the more traditional (110°—24 hr) procedure for a diverse set of sample types. The results demonstrate the feasibility of the higher temperature, shorter hydrolysis time for many applications. The results of a prehydrolysis oxidation for analysis of cystine as cysteic acid and methionine as methionine sulfone for the diverse samples were evaluated, and amino acid values extrapolated from multiple hydrolysis times at 145° were compared with multiple hydrolysis at 110°, illustrating a means for obtaining values for amino acids such as isoleucine, valine, threonine, and serine which are more accurate than those from a single hydrolysate, and more rapid than the typical 110°-multiple hydrolysate technique.

We also evaluated interlaboratory variations in sample preparation, which revealed that although differences resulting from hydrolysate preparation by two different laboratories can be minimized, those variations are greater than the chromatographic or analytical variability. These studies further demonstrate that use of glass tubes with Teflon®-lined screw caps as hydrolysis vessels compares favorably with sealed glass ampules and that 145°—4 hr hydrolysis with these tubes, after careful exclusion of air, is a rapid, precise, and practical method for protein hydrolysis. The importance of careful sample preparation to the achievement of accurate data is stressed in Chapter 1, which also provides both a review of HCl hydrolysis of proteins and practical information on preparation of hydrolysates.

Chapter 2

The GC analysis of amino acids as the N-trifluoroacetyl (N-TFA) n-butyl esters is the subject of Chapter 2. This established method, principally developed in the author-editors' laboratories at the University of Missouri-Columbia, provides an effective and reliable means of amino acid determination that is applicable to a very wide range of analytical needs. Prof. Gehrke, graduate students, and colleagues during the period from 1962 to 1975 established the fundamentals of quantitative derivatization, conditions of chromatographic separation, and defined the interactions of the amino acid derivatives with the stationary and support phases. Their studies and continued refinements since 1975 have resulted in a precise and accurate, reliable, and straightforward method for amino acid measurement.

The chapter begins with an extensive review of the applications of GC to amino acid analysis of a wide range of matrices, from pine needle extracts to erythrocytes. The *Experimental* section provides a thorough description of the quantitative analytical procedures developed by the authors, including preparation of the ethylene glycol adipate (EGA) and silicone mixed phase chromatographic columns. The EGA column which is used to separate and quantitate all the protein amino acids except histidine, arginine, and cystine is composed of 0.65 w/w% stabilized grade EGA on 80/100 mesh acid-washed Chromosorb® W, 1.5 m × 4 mm I.D. glass. For quantitation of histidine, arginine, and cystine the silicone mixed phase of 1.0 w/w% OV-7 and 0.75 w/w% SP-2401 on 100/120 mesh Gas-Chrom® Q (1.5 m × 4 mm I.D. glass) performs extremely well. They also describe the preparation and use of ion-exchange resins for sample cleanup, and complete sample derivatization to the N-TFA n-butyl esters. The amino acids are esterified by reaction with n-butanol · 3 N HCl for 15 min at 100°C, the excess n-butanol · 3 N HCl removed under vacuum at 60°C, any remaining moisture removed azeotropically with dichloromethane, then the amino acid esters are trifluoroacylated by reaction with trifluoroacetic anhydride (TFAA) at 150°C — 5 min in the presence of dichloromethane as solvent. Immediately following the *Experimental* section are valuable comments on various parts of the method which provide guidance to the use of the entire technique, from sample preparation to chromatography to quantitation.

Of particular value in this chapter is the comparison of GLC and IEC results of hydrolysates of diverse matrices. This extensive comparison of an array of sample types showed that the values obtained by the two techniques were generally in close agreement.

GLC and IEC analyses of multiple hydrolysates were performed to evaluate the reproducibility of hydrolysate preparation and to compare GLC and IEC analyses of the same hydrolysates. The total amino acids found in the same hydrolysates were essentially identical by both GLC vs. IEC. As the sets of three hydrolysates were prepared at the same time under identical conditions, it might be expected that differences between the GLC and IEC analyses of the same hydrolysates would be greater than the differences between identically prepared hydrolysates. However, slight differences in the amounts of certain amino acids present in the different hydrolysates can be observed, emphasizing that variations do arise due to the hydrolysis itself, even under preparation conditions most conducive to reproducibility.

As the sulfur-containing amino acids are of particular interest in nutrition, cystine and methionine analyses are discussed in detail. The quantitative determination of amino acids in addition to cystine and methionine in preoxidized hydrolysates by IEC is described, and a rapid oxidation-hydrolysis procedure is presented which allows accurate analysis of cystine, methionine, lysine, and nine other amino acids in feedstuffs and other biological matrices.

One of the authors of Chapter 2 (FEK) has used the *N*-TFA *n*-butyl ester method for more than 17 years on a routine basis in a commercial laboratory (Analytical Biochemistry Laboratories, Columbia, Mo.) and his observations on the analysis of an extremely wide range of sample types over this time span are presented in a special section of this chapter. *Experiences of a Commercial Laboratory* provides valuable practical information into amino acid analysis by GC.

The analysis of amino acids as the *N*-TFA *n*-butyl esters is an established technique that offers much to scientists concerned with the determination of amino acids. The method offers excellent precision, accuracy, selectivity, and is an economical complementary technique to the elegant Stein-Moore ion-exchange method.

Chapter 3

In this chapter, we provide both a detailed account and historical perspective of our development of GC amino acid analysis, and describe the solution of problems encountered as the methods evolved. The *N*-TFA *n*-butyl ester and trimethylsilyl (TMS) derivatives are discussed, including reaction conditions, chromatographic separations, mass spectrometric (MS) identification of both classes of derivatives, interactions of the arginine, histidine, and cystine derivatives with liquid phase and support materials, and application of the methods.

The acylation of arginine posed a problem in early studies, and the successful solution of this particular problem paved the way to a high-temperature acylation procedure which is now widely used with numerous acylating reagents. Likewise, esterification of the amino acids was investigated in detail, resulting in a direct esterification procedure which quickly and reproducibly converts the amino acids to the *n*-butyl esters. This approach has also been widely used to form various amino acid esters.

Chapter 3 describes the development of chromatographic columns for separation of the *N*-TFA *n*-butyl esters, from early efforts to obtain a single column separation of the protein amino acids to the successful development of a dual column system which has been used on a routine basis for some 15 years for quantitative amino acid analysis.

The early development of GLC analysis of iodine- and sulfur-containing amino acids as the TMS derivatives is described, with the finding that bis(trimethylsilyl)-trifluoroacetamide

(BSTFA), a silylating reagent which we invented, is an effective silylating reagent for forming amino acid derivatives.

Our studies on the derivatization of the protein amino acids with our new reagent, BSTFA, is described in which the amino acids are converted to volatile derivatives in a single reaction step. Although certain amino acids tend to form multiple derivatives which contain varying numbers of TMS groups, high-temperature, long reaction time derivatization permits quantitative analysis of the amino acids as the TMS derivatives. Our studies on the GLC separation of the TMS amino acids resulted in the development of a 6 m column of 10% OV-11 on Supelcoport® for separation of the TMS derivatives.

The development of a chromatographic column system for the *N*-TFA *n*-butyl esters came about from the realization that the derivatives of arginine, histidine, and cystine were not reproducibly eluted from columns with polyester liquid phases, although this type column was excellent for analysis of the other protein amino acids. We developed a siloxane mixed phase column specifically for these 3 amino acids, with the developed system being an EGA for 17 amino acids and the mixed phase column for these remaining 3. Studies on derivative interactions with liquid phases and support materials confirmed that the polar liquid phase EGA was primarily responsible for the destruction of the histidine derivative, and that the derivatives of arginine, histidine, and cystine are all subject to complex temperature-dependent interactions with EGA and nondeactivated support material.

Our invention of a GC solvent venting system is also described in Chapter 3. The venting system allows injection of 100 μℓ or more, but prevents the solvent and excess acylating reagent from traversing the column while allowing quantitative transport of the amino acid derivatives through the column to the detector. This system was invented as a direct result of our involvement in the search for amino acids in the returned lunar samples, for which we wished to increase the volume of sample injected in order to search for part-per-billion amounts of amino acids. At that same time in the lunar studies we also conducted a survey of potential sources of low-level amino acid contamination which is also described in Chapter 3.

The power of GC/MS for identifying amino acids in complex matrices prompted us to obtain and publish the electron-impact mass spectra of the *N*-TFA *n*-butyl ester and TMS derivatives of the amino acids. We obtained mass spectra of 48 *N*-TFA *n*-butyl amino acids and 46 TMS amino acid derivatives, and Chapter 3 summarizes the major characteristics of their spectra.

Chapter 3 ends with a summary which points out that the foundation of successful amino acid analysis by GC is composed of two elements: (1) reproducible and quantitative conversion of amino acids to suitable derivatives and (2) separation and quantitative elution of the derivatives by the chromatographic column. This chapter provides the reader with valuable background information into the development and successful use of GC for amino acid analysis.

Chapter 4

S. L. MacKenzie of the Plant Biotechnology Institute, NRC, Canada, a leader in the successful development of the *N*-heptafluorobutyryl (*N*-HFB) isobutyl ester derivatives, describes the rationale leading to his extensive work, and he presents in detail the derivatization, separation and applications of this derivative in Chapter 4. His chapter contains a section entitled *Important Comments,* pointing out that derivatization is the most crucial factor in reproducible analysis of amino acids by GLC. His observations in that section will be particularly helpful to the analyst interested in utilizing the *N*-HFB isobutyl ester method. MacKenzie then discusses the analysis of protein hydrolysates and free amino acids, reviews the analysis of nonprotein amino acids and presents interesting applications of the analysis of free amino acids, such as studies of the free amino acid pool of starfish tissue, and whether sulfur amino acids are involved in the mercury resistance of certain fungi. Chapter 4 concludes with an informative section on electron impact and chemical ionization mass spectral analysis of the *N*-HFB amino acid isobutyl esters.

Chapter 5

Noting that there are some 50 diseases known to be due to anomalies of amino acid metabolism, J. Desgres and P. Padieu of the National Center for Mass Spectrometry and the Laboratory of Medical Biochemistry at the University of Dijon, France, describe the development of the N-HFB isobutyl derivatives for the clinical analysis of amino acids, and the adaptation of the method to the routine practice of a clinical biochemistry laboratory in Chapter 5. The experimental protocol is described in detail, followed by application of the procedure for analysis of normal and pathologic physiological fluids including phenylketonuria, maple syrup disease, idiopathic glycinemia, cystathionase deficiency, cystathionase synthetase deficiency, and renal absorption disorders. Desgres and Padieu have routinely used their method for more than 6 years in a clinical biology laboratory, and conclude that GLC is perfectly suited for the daily analysis of more than 30 amino acids; they also point out the importance of GC/MS in elucidating metabolic disorders.

Chapter 6

I. Moodie of the National Research Institute for Nutritional Diseases and the Metabolic Unit, Tygerberg Hospital, South Africa, describes in Chapter 6 his development of an efficient GLC method specifically for routinely producing accurate determination of protein amino acids in fishery products. Moodie presents a thorough account of his comprehensive studies, describing his choice of suitable derivative and columns, modification of derivative preparation, and sample preparation techniques for packed column analysis of the *N*-HFB isobutyl esters. He describes his chromatographic procedure in detail, then reports the use of the procedure to statistically study the effects of sampling and hydrolysis on the precision of analysis. Moodie finds that a major contribution to reduced precision arises during sampling and hydrolysis. Moodie has extensively studied glass capillary columns for amino acid analysis and describes column preparation and use, column longevity, and comparisons of packed and capillary columns. His chapter serves as an excellent example of the practical utilization of amino acid analysis in nutrition by GLC.

VOLUME II — INTRODUCTION

Chapter 1

Development of the chiral diamide phases for resolution of amino acid enantiomers is the subject of Chapter 1 by E. Gil-Av, R. Charles, and S.-C. Chang of the Weizmann Institute of Science, Israel.

Pioneers in research on the separation of enantiomers, Gil-Av et al. discuss the evolution of optically active phases from the α-amino acid derivative phases (e.g., N-TFA-L-Ile-lauroyl ester) to dipeptide phases (e.g., N-TFA-L-Val-L-Val-O-cyclohexyl) to the still more efficient and versatile diamide phases of the formula $R_1CONHCH(R_2)CONHR_3$. Gil-Av et al. describe the synthesis and purification of the diamide phases, and the determination of their optical purity.

The influence of structural features of the diamides on resolution and thermal stability is discussed in detail. In these extensive studies, the structures of R_1, R_2, and R_3 in the above formula were varied with R_1 and R_3 representing various n-alkyl, branched alkyl, and alicyclic groups, and R_2 representing various aliphatic and aromatic groups. Chain lengthening of R_1 and R_3 groups produced the desired increase in thermal stability, yielding phases operable at 200°C and above.

Gil-Av then follows with an elegant discussion of the mechanism of resolution of amino acid enantiomers with the diamide phases, using models based on the consideration of conformations and modes of hydrogen bonding of simple diamides.

The effect of N-acyl and ester groups on chromatographic separation is described, with the conclusion that isopropyl esters are the best compromise with regard to volatility, resolution, and ease of preparation. Differences among the N-trifluoroacetyl (N-TFA), N-pentafluoropropionyl (N-PFP), and N-heptafluorobutyryl (N-HFB) groups were reported to be small, but perhaps could be exploited to overcome problems of peak overlap of amino acid derivatives emerging very close to each other. Gil-Av also presents information on the influence of the side chain attached to the asymmetric carbon of the amino acids on resolution.

This chapter provides an excellent description of enantiomeric analyses with diamide phases, with discussions of analyses of samples without and with peak overlap, and the use of phases with opposite configuration for identification purposes. In conclusion, Gil-Av et al. compare GLC and HPLC for resolution of amino acid enantiomers, and notes the advantages of the GLC techniques.

Chapter 2

The separation of amino acid enantiomers using chiral polysiloxane liquid phases and quantitative analysis by the novel approach of "enantiomeric labeling" is presented by E. Bayer, G. Nicholson, and H. Frank of the University of Tübingen, West Germany. The synthesis of a novel type of chiral stationary phase was accomplished by Bayer et al. which possesses sufficiently high thermal stability to enable GC separation of the enantiomers of all the protein amino acids within about 30 min. The stationary phase, subsequently named Chirasil-Val®, is a polysiloxane functionalized with carboxy alkyl groups to which valine-t-butylamide residues are coupled through and amide linkage. Bayer et al. describe the criteria which the stationary phase must fulfill, and discuss the steps taken to synthesize the stationary phase to meet those requirements. They also describe the preparation of capillary columns with Chirasil-Val® stationary phases and GC conditions needed to optimize the separations.

Bayer et al. studied the effects of the acyl and ester groups on the "enantiomeric resolution" and relative retention times of each amino acid on Chirasil-Val®, and concluded that for complete resolution of all protein amino acids, a number of acyl ester combinations are possible, notably the N-PFP n-propyl esters and isopropyl esters, and the N-TFA n-propyl esters.

The concept of the use of D-amino acids as internal standards for the quantitative analysis of the L-amino acids is first and elegantly described in Chapter 2. Bayer et al. point out the prior to analyses of amino acids in biological fluids, a procedure comprised of several steps is usually required to separate the amino acids from high molecular weight material, carbohydrate derivatives, fatty acids, oligopeptides, biogenic amines, etc., and that recovery of the amino acids during this process must be either quantitative or of a constant and known percentage. Also, variations can occur during derivatization and in the response of the detector, and discriminatory effects in the injector can arise when using the split technique of sample injection.

By using the optical antipodes as internal standards, not only are volumetric losses during work-up, derivatization and GC compensated, but also differences in chemical behavior such as recovery from ion-exchange columns during cleanup, yield of derivatization, and stability of the derivatives. With "enantiomeric labeling", the internal standard suffers exactly the same fate as its optical antipode. Bayer et al. illustrate the use of "enantiomeric labeling" for analyses of blood serum and gelatin, and present the mathematical expressions for determining the concentration of each amino acid in samples. Sensitivity, precision, and accuracy of the enantiomeric labeling technique are discussed, with accuracy of determination of synthetic samples surpassing ion-exchange chromatography (IEC). Thus GC analysis of amino acids by "enantiomeric labeling" is a powerful technique for a wide variety of applications. When many analyses are required Bayer also has introduced a new automated derivatization robot.

Chapter 3

The analysis of amino acid racemization and the separation of amino acid enantiomers is presented by H. Frank of the University of Tübingen, West Germany. The author presents examples of enantiomers of free amino acids and amino acid analogues which exhibit different biological properties. A tragic illustration is the glutamic acid derivative thalidomide, a sedative drug used in the 1950s which induced malformations of newborns. The L-enantiomer of a metabolite is teratogenic, the D-antipode is not.

The importance of stereochemical analysis of amino acids for studies on the biochemical, pharmacological, or toxicological properties of anantiomers is described, as well as its utilization in structure elucidation, peptide synthesis, racemization monitoring, and pharmaceutical research.

Frank discusses the assessment of racemization in peptide synthesis by separation of dipeptide diastereomers, detailing factors which give rise to racemization during peptide synthesis and describing how racemization can be suppressed. Then follows a comprehensive discussion of mechanisms involved in amino acid racemization, including factors such as the chemical state (free amino acid, derivative, or in peptide linkage) temperature and pH.

The chromatographic measurement of the D- and L-enantiomers as diastereomers and directly as enantiomers is discussed and compared, including factors which introduce error into the analysis. The use of GC/chemical ionization-mass spectrometry (CI-MS) in racemization studies is described in which deuterium/hydrogen exchange during hydrolysis is monitored by GC/CI-MS to ascertain the extent of incorporation of deuterium into each enantiomer.

Frank presents a thorough discussion of the mechanisms involved in hydrogen/deuterium exchange, which is a powerful tool for determination of true enantiomer ratios in synthetic peptides and proteins. Detailed analysis of racemization mechanisms is possible by this procedure, and in many cases it represents the sole possibility to discriminate between racemate generated during hydrolysis and the amount originally present.

Deuterium/hydrogen exchange and GC/MS allow measurement of much lower amounts of D-enantiomers in proteins than GC alone. Using only GC, a lower limit of measurement is about 0.1% of the D-enantiomer in the presence of 99.9% of the L-enantiomer. This limit is a result of both the linear range of the chromatographic system (assuming a practical range of three orders of magnitude) and the fact that all amino acids form 0.1% or more of D-enantiomer during hydrolysis. For meaningful determination of lower percentages of D-amino acids in proteins one has to resort to deuterium/hydrogen exchange and GC/MS which enables the determination of 0.01% of D-enantiomer originally present in a protein.

Frank describes a technique to measure even smaller quantities of D-enantiomers by use of an added standard compound as a reference peak. By using a series of dilutions, ideally D-enantiomers could be measured when present at 0.01 to 0.001% of the L-enantiomer. The methods for determination of optical purity of amino acids have been greatly refined, making previously intractable problems now appear relatively straightforward.

Chapter 4

A. Darbre, of Kings College, London, one of the first investigators of the use of GC to determine amino acids, includes an interesting historical perspective in Chapter 4. He describes studies of the N-TFA amyl and methyl esters, including column evaluations, various methods of esterification, trifluoroacylation, and quantitative studies with gas density balance and flame ionization detectors. His extensive investigations on the chromatography of amino acid derivatives includes use of radioactive amino acids to study the breakdown of some derivatives during chromatography by monitoring of the column effluent for radioactivity. Darbre also describes his studies on methods of solid sample injection, *in situ* precolumn derivatization, and quantitative studies of biological samples using the N-TFA amino acid methyl esters. He also discussed the GC separation of acylated dipeptide methyl esters, pointing out the recent advances in MS techniques for peptide analysis.

Darbre's most recent work has centered on the N-HFB isobutyl esters, and he describes a practical derivatization method and its application to analysis of blood plasma and other biological samples. The use of electron-capture (ECD), alkali flame (AFID), and flame-ionization detection (FID) are also treated in this chapter, along with comparisons of relative molar response (RMR) values of the amino acid derivatives and the analysis of biological samples by capillary columns with FID, AFID, and ECD. Darbre foresees the next stage of development to include work at even higher levels of sensitivity (nanomole to picomole amounts), and perhaps eventually the use of aseptic conditions to eliminate the problems of background contamination.

Chapter 5

The N-TFA n-propyl esters have been the subject of an in depth study by G. Gamerith, at the University of Graz, Austria, and his work on the development of this technique and derivative is described in Chapter 5 for medical diagnostic applications. Gamerith thoroughly discusses his effective amino acid analysis method for screening individuals with potential inborn errors of metabolism, and demonstrates the application of the method for quantitative screening of urine specimens of patients with mental retardation.

Chapter 6

Cyril Ponnamperuma of the University of Maryland elegantly presents the role of GC in the study of cosmochemistry in Chapter 6. One of the most fundamental questions of all science is the "origin of life". How did it first begin? Since the laws of physics and chemistry are universal laws it may be legitimate to extrapolate from the circumstances on earth to elsewhere in the universe, and our present day telescopes reveal that there are 10^{23} stars which like our own sun can provide the photochemical basis for plant and animal life.

To the Russian biochemist, A. I. Oparin, more than to anyone else today, we owe our present ideas on the scientific approach to the study of the origin of life. In clear and scientifically defensible terms he pointed out that the complex combination of manifestations and properties so characteristic of life must have arisen in the process of the evolution of matter. It was in 1924 that he enunciated his hypothesis on the "Chemical Evolution of Life Molecules."

From this date it was left to the analytical biochemist to develop the powerful research tools of chromatography that enabled the chemist to unravel a problem awesome by its very nature and importance. This chapter presents the central role of GC and interfaced MS spanning 2 decades of study on chemical evolution, prebiotic chemistry, our search for life molecules on the moon and planets, ancient sediments, and the extraterrestrial Murchison Meteorite. The detection and quantitation of minute traces of amino acids in complex mixtures was made possible by the highly sensitive and selective GC techniques that were first developed at the University of Missouri by Gehrke, Kuo, and Zumwalt, and which are described in detail in these volumes.

Thus, with the data of our analyses, we were presented with the unequivocal conclusion that the presence and survival of amino acids at the 10^{-9} g/g level in the environment of the lunar soil appears to be highly unlikely. However, our experiments in 1984, with the methods of HPLC followed by capillary GC interfaced with MS presented for the first time unambigious evidence for the existence of extraterrestial nucleobases in the Murchison meteorite. The nucleobases uracil, thymine, cytosine, ademine, and xanthine were found in the primordial soup generated by an electric discharge and also in the Murchison Meteorite. GC/MS has thus provided the chemist with a powerful research tool for the study of organic cosmochemistry.

Chapter 7

Volume II ends with a description by the author-editors of one of the most unusual applications of amino acid analysis by GC. Our 5-year search from 1969 to 1974 for amino acids on the 4.5 billion-year-old returned lunar samples was conducted at the NASA Ames Research Center, California, University of Missouri, Columbia, and at the Laboratory for Chemical Evolution, University of Maryland. This was an exciting research period, as these investigations could result in findings which would advance our knowledge of the processes of chemical evolution, and thus the Apollo missions have provided us with the requisite extraterrestrial material for these studies. Our primary aim throughout the investigations of the return lunar samples was the search for, and identification of, water-extractable, derivatizable organic compounds with particular emphasis on the amino acids. An interesting comparison of experimental findings was made in 1971 by two different methods of amino acid analysis using GLC and IEC. This special collaborative study was organized by NASA and conducted on lunar materials obtained by the Apollo 14 mission. The objective was to ascertain whether amino acids were present in the "free" (F) and "hydrolyzed" (H) water extracts of the lunar fines by comparison of GLC and IEC results. The GC analyses were made by Gehrke, Kuo, Zumwalt, and Ponnamperuma while Hare and Harada used ion-exchange methods. The two complementary methods gave closely similar and, therefore, confirmatory analytical results when applied to samples which had been processed and handled in identical ways by the two groups of researchers.

Our analyses by both GLC and IEC on the hot water sample extracts produced independent yet supporting analytical results, with the unequivocal conclusion from our 5 years of study that the presence and survival of amino acids at 10^{-9} g/g level in the environment of the lunar regolith appears to be highly unlikely.

A powerful "research tool" has been provided in the GC methods presented by the authors in these three volumes for the measurement of amino acids in sample types ranging from pure proteins to biological and geological materials. These methods will open the doors to the solutions of many problems in science.

VOLUME III — INTRODUCTION

Chapter 1

W. A. König of the University of Hamburg, West Germany, describes the analysis of amino acids and *N*-methylamino acids by capillary GC of diastereomeric derivatives on nonchiral phases and the use of isocynates as reagents for enantiomer separation on chiral phases in Chapter 1. König discusses practical applications of configurational analysis of amino acids, and presents in detail the use of diastereomeric derivatives. Factors which affect the precision of determinations of the proportion of a mixture of enantiomers are described, including optical purity of the chiral reagent and reaction kinetics of enantiomers in the formation of diastereomeric derivatives.

The two routes for obtaining diastereomeric derivatives of amino acids are presented; esterification with a chiral alcohol or use of the amino function for conversion of amino acids to diastereomeric derivatives. König describes the preparation of the chiral alcohol (+)-3-methyl-2-butanol, esterification of amino acids with that alcohol to form diastereomers, followed by acylation of the amino group to PFP or TFA derivatives. Utilization of the diastereomeric technique is described for unusual and *N*-methylamino acids which are commonly found in many peptide antibiotics, as well as applications to α-hydroxy acids and α-branched carboxylic acids.

Chapter 2

Abe and Kuramoto address in depth two topics in their chapter; first, a comparison of the resolution of various D,L-amino acid derivatives on optically active columns, and second, the "age dating" of sediments by determination of the extent of racemization of amino acids in fossil shells.

The influence of the ester group on chromatographic resolution was evaluated by synthesis and analysis of the methyl, ethyl, *n*-propyl, isopropyl, and neopentyl esters, and the effect of the acyl group was investigated by preparation and analysis of the TFA, PFP, and HFB derivatives. A total of nine different derivatives were studied and the effects of the different substituent groups were investigated in terms of retention times, D and L peak-to-peak resolution, and elution order on both glass and fused silica Chirasil-Val® columns. This systematic study provides detailed information on the separation of amino acid enantiomers and Abe points out that he experienced greatest difficulty in separation of the lower molecular weight amino acids (glycine, threonine, isoleucine, *allo*isoleucine) which elute within a very narrow range.

Abe and Kuramoto describe the application of enantiomeric analysis to age dating of sediments of the Osaka Plain. The kinetics of amino acid racemization are described with the rate being dependent on temperature, pH, and ionic strength. Temperature especially affects the rate of racemization, e.g., racemization half-lives at 0°C are about 100 times longer than at 25°C. The expressions for calculation of the age of fossils from the enantiomer ratios are then presented. The age of at least one fossil from the region is determined by radiocarbon dating and that sample is then used as a calibration sample for determining the age of other fossils. Sample cleaning, hydrolysis, ion-exchange purification, and derivatization and analysis are discussed, and a very good *Comments on the Method* section is included in the chapter.

Sediments from 50 and 64.5 m in depth were dated at 27,000 to 28,000 and 29,000 to 30,000 years before present, respectively, illustrating the potential for age dating from D,L ratios of several amino acids.

Comparison of glass and fused silica Chirasil-Val® columns showed the glass columns yielded better resolution, but the fused silica columns possessed superior thermal stability. This chapter provides the reader with both comparative chromatographic data on nine different derivatives and provides an excellent illustration of the potential of configurational analysis to address a specific question of age dating.

Chapter 3

N. Ôi, of Sumika Chemical Analysis Ltd., Osaka, Japan, describes his extensive research on the use of S-triazine derivatives of amino acid esters, peptide esters, and amino acid amides as chiral stationary phases for separation of amino acid enantiomers. Ôi initially found that the S-triazine structure provided improved thermal stability while these derivatives of amino acid esters exhibited good enantioselectivity. He proceeded with the synthesis and evaluation of S-triazine derivatives of peptide esters as stationary phases, comparing dipeptide and tripeptide derivatives which possess greater maximum operating temperatures than the amino acid ester derivatives. Turning to the amino acid amides, Ôi synthesized an S-triazine derivative that exhibited even better enantioselectivity than the peptide ester phases and which allowed separation of some enantiomers on packed columns.

Ôi also emphasizes that the S-triazine derivatives are effective for anantiomeric separation of other classes of organic compounds. He reports the successful resolution of the enantiomers of some linear and cyclic peptides, amines, α-phenylcarboxylic acids, α-hydroxycarboxylic acids, alcohols, and an organophosphorus compound which contains an asymmetric phosphorous but no asymmetric carbon atom.

Ôi describes the determination of the optical purity of proline, which is important in research when proline is used as a chiral reagent in the form of an N-acyl chloride for formation of diastereomers. As the enantiomers of proline are more difficult to separate than other amino acid enantiomers, Ôi developed and describes the excellent separation of proline enantiomers in the form of N-acylisopropylamides on S-triazine derivative phases. Similarly, he describes the enantiomeric analysis of 1-phenylethylamine, another chiral reagent used for separation of enantiomeric pairs of various chiral acids.

Ôi demonstrates the analysis of enantiomers of a constituent of a synthetic pyrethroid which has been developed as an agricultural pest control agent, and the enantiomeric analysis of allethrolone which is an important constituent of synthetic insecticidal pyrethroids.

Ôi concludes that the novel S-triazine derivatives of amino acid esters, peptide esters, and amino acid amides are effective stationary phases for configurational analysis of a range of enantiomeric compounds.

Chapter 4

G. Odham and G. Bengtsson of the University of Lund, Sweden, present an excellent description of the GC microanalysis of amino acids using electron capture detection (ECD) and MS with emphasis on microenvironmental applications, especially in studies on the interactions between microorganisms and other cells. Their research has focused on situations in which the sample size is limited to microliters or micrograms, and the amino acid concentration are in the ppb or ppt range. This combined limitation of both very low sample amount and very low concentrations within the sample creates special analytical demands which the authors clearly address.

They describe micro-scale sample preparation including hydrolysis (for proteins and peptides), extraction (for free amino acids), ion-exchange cleanup, and derivatization, providing much practical information on each procedure. For capillary GC separation, their use of splitless and on-column injections is described as limited sample quantity prohibited use of the split technique. The authors point out that both splitless and on-column injections permit quantitative analysis of amino acids, while split injection is unreliable at least for the higher boiling amino acids.

Odham and Bengtsson discuss the advantages and limitations of support-coated and wall-coated open tubular columns for microanalysis, noting that narrow-bore (0.2 to 0.3 mm I.D.) wall-coated columns provide the best separations. They discuss the use of capillary columns with chiral stationary phases for study of cellular interactions involving prokaryotic bacteria as these organisms contain unique amino acid enantiomers such as D-alanine and D-glutamic acid in defined proportions in their cell walls. They illustrate the GC/MS determination of the D,L ratio of alanine in intact cells of Group A *streptococci* type A using a chiral phase capillary column.

The authors provide a thorough discussion of GC detection and quantitation of amino acids as the HFB isobutyl derivatives. EC detectors allow detection of less than 1 pg of amino acid HFB isobutyl ester, and the authors reported the linearity was satisfactory within the 10 to 400 pg range.

Odham and Bengtsson describe GC/quadrupole MS for microanalysis of amino acids, illustrating the use of repetitive scanning and extracted ion current profiles for analysis of microgram quantities of *Myxococcus virenscens*. They discuss selected ion monitoring as a means of increasing analytical sensitivity and report that the detection limit for valine was determined by the general background level of the laboratory, 0.8 pg. The authors provide a detailed account of quantitation by MS, discussing various internal standard compounds such as those structurally similar to the amino acids, (e.g., norleucine), the optical antipodes (D-amino acids), and the optimal stable-isotope-labeled analogue (amino acids labeled with ^2H, ^{13}C, or ^{15}N). The isotope label technique is shown to offer excellent precision of measurement and the authors offer a clear perspective of factors which influence the sensitivity that can be achieved in the quantitative MS of amino acids. The authors show that measurment of the $^{15}N/^{14}N$ ratios in amino acids, using selected ion monitoring of major *N*-containing fragments, can provide an accurate means of monitoring nitrogen metabolism in biological systems.

Odham and Bengtsson have applied their research to both ecological and medical problems. For example, their technique allowed quantitation of amino acids in single dew droplets (50 $\mu\ell$) on leaves and the demonstration of mycoplasma contamination of tissue cell cultures by measurement of ornithine exuded by *Mycoplasma hominis*.

In conclusion, the methods of Odham and Bengtsson for the ultramicrodetermination of amino acids reported in Chapter 4 extend amino acid analysis into microenvironments (e.g., a single aphid on a leaf). Their work opens new possibilities to study microbial interactions in a variety of microenvironments and the applications described illustrate the potential of the micromethod in ecological and medical research.

Chapter 5

The oxazolidinonees are the subject of Chapter 5. P. Hušek of the Endocrinology Institute, Prague, Czechoslovakia, has developed a novel derivative, and in his chapter, Hušek considers alternatives to the esterification-acylation approach to amino acid derivatization and presents possibilities for the simultaneous derivatization of β-amino and carboxyl groups. He investigated the use of perhalogenated acetones, i.e., 1,3-dichloro-1,1,3,3,-tetrafluoroacetone (DCTFA), for forming cyclic derivatives thus developing a "single medium, two-reagent" reaction approach for amino acid derivatization. Amino acids which do not contain side-chain reactive groups can be analyzed after cyclization only, while those with reactive side-chain groups are analyzed after treatment with an acylating agent. Hušek clearly describes the derivatization procedures and chromatographic analysis after cyclization and acylation. He also discusses applications of his analytical method to analysis of physiological fluids, recommends a practical procedure for isolation of free amino acids from 50 to 100 $\mu\ell$ of serum or urine, and presents a rapid procedure for the determination of thyroid hormonal compounds, diiodothyronine, triiodothyronine, thyroxine, and their deamination and decarboxylation products.

Chapter 6

M. Makita and S. Yamamoto at Okayama University, Japan, have led the development of the *N*-isobutoxycarbonyl (isoBOC) methyl ester derivatives of the amino acids and describe their research in Chapter 6. This approach offers a potentially rapid and quantitative means of amino acid derivatization (without heating), which yields derivatives stable to moisture. Makita and Yamamoto present the procedures for derivative and column preparation, provide helpful comments on procedures, and summarize the GC/MS structure elucidation of the derivatives. GC separation studies of the *N*-isoBOC methyl esters are described, including the separate analysis of arginine as ornithine after treatment of the sample with arginase, as are the quantitative aspects of the entire method. The analytical applications presented include analysis of protein hydrolysates, serum amino acids, and determination of hydroxyproline in urine as an indicator of collagen metabolism. A method is also described for the analysis of urinary γ-carboxyglutamic acid. This amino acid, present in several calcium-binding proteins, appears in urine as a degradation product of these proteins and is a potential indicator for assessment of diseases associated with blood coagulation and bone metabolism. The mild derivatization conditions permit GC analysis of γ-carboxyglutamic acid, as this amino acid is highly acid-labile, being converted to glutamic acid. In their chapter, Makita and Yamamoto demonstrate the GC analysis of 22 protein amino acids, including asparagine and glutamine, as the isoBOC methyl ester derivatives.

Chapter 7

Capillary GC, invented by M. J. E. Golay, was first described in 1957. Until recently, capillary columns were mainly made of glass or stainless steel, but with the introduction of fused silica columns in 1979 by Dandenau and Zerenner, fused silica has largely replaced all other column materials.

Techniques for coating the inside wall of the column tubing with liquid phase were greatly improved with the advent of ''immobilized'' liquid phases. These phases are created by polymerizing a prepolymer inside the column, resulting in a high molecular weight, cross-linked phase that is also bonded to the Si-OH groups of the fused silica surface. The product of these developments are flexible, inert, mechanically durable, highly reproducible, and efficient columns which can be operated over a wide temperature range. These characteristics were viewed by us as being extremely promising for analysis of amino acid derivatives, and studies were undertaken to evaluate this type of column for one of the most often-sought determinations in protein research: the measurement of amino acids in hydrolysates of proteinaceous materials.

This study took advantage of the fact that two different derivative types have been thoroughly investigated and developed, and are now well-established for packed column amino acid analysis: the *N*-TFA *n*-butyl esters developed by Gehrke et al. (Chapters 1 to 3, Volume I) and the *N*-HFB isobutyl derivatives primarily developed by MacKenzie and coworkers (Chapter 4, Volume I).

In Chapter 7, we have demonstrated that fused silica ''bonded phase'' columns are a very effective means for separating and quantitatively measuring amino acids in protein hydrolysates as both the *N*-TFA *n*-butyl and *N*-HFB isobutyl derivatives. Comparisons are presented of columns with differing liquid phase polarities, providing information on the effect of polarity on separation of particular amino acids, and evaluation of split, splitless, and on-column injection methods revealed the on-column injection technique to be most reproducible.

Fused silica capillary columns with chemically bonded phases of 1.0 μm film thickness (J & W Durabond DB-1 and DB-5) were used to separate and quantitate the amino acid derivatives. All the protein amino acid *N*-TFA *n*-butyl esters were separated on a short (20 m) DB-5 column, while only the derivatives of lysine and tyrosine were not well resolved on a DB-1 column of the same length. Conversely, the DB-5 column gave better separation of all the protein amino acid HFB isobutyl esters than did the DB-1 column, with the DB-5 column providing an improved lysine/tyrosine separation.

A number of analytical applications on diverse sample types are presented in which data from fused silica capillary GC analyses are compared to results obtained form IEC.

Amino acid analysis of commercial animal feeds, feed ingredients, lysozyme, β-lacto-globulin, and ribonuclease by capillary GLC as both the N-TFA n-butyl esters and the N-HFB isobutyl esters were in agreement with IEC results. The comparisons clearly demonstrate that the capillary GC method is a powerful tool for the measurement of amino acids. In summary, we describe the uniting of established methods of amino acid derivatization which have been developed by Gehrke et al. and MacKenzie with the recent development of fused silica immobilized phase capillary columns into a high resolution, precise, and accurate method for measurement of amino acids.

CONTRIBUTORS

Iwao Abe, Ph.D.
Research Assistant
Department of Applied Chemistry
College of Engineering
University of Osaka Prefecture
Osaka, Japan

Joseph S. Absheer, M.Sc.
Research Chemist
Experiment Station Chemical Labs
University of Missouri-Columbia
Columbia, Missouri

Ernst Bayer, Ph.D.
Professor and Director
Institut für Organische Chemie
Universität Tübingen
Tübingen, West Germany

Göran Bengtsson, Ph.D.
Docent-Research Associate Professor
Laboratory of Ecological Chemistry
Ecology Building
University of Lund
Lund, Sweden

Shu-Cheng Chang, Ph.D.
Department of Chemistry
Chung-Shan Institute of Science and
 Technology
Lung-Tan, Taiwan

Rosita Charles, M.Sc.
Research Chemist
Department of Organic Chemistry
Weizmann Institute of Science
Rehovot, Israel

André Darbre, Ph.D.
Senior Lecturer, Adviser of Studies
Biochemistry Department
King's College London
Strand, London, England

Jean Desgrès, Ph.D.
Assistant Professor
Department of Medical
 Biochemistry
Faculté de Médecine
Centre Hospitalier
 Régional Universitaire
Université de Dijon
Dijon, France

Hartmut Frank, Ph.D.
Institut für Toxikologie
Universität Tübingen
Tübingen, West Germany

Gernot Gamerith, Ph.D.
Department of Research
 and Development
Lenzing AG
Lenzing, Austria

Charles W. Gehrke, Ph.D.
Professor of Biochemistry and Director
 Interdisciplinary Chromatography-Mass
 Spectrometry Facility
Department of Biochemistry
University of Missouri-Columbia
Columbia, Missouri

Emanual Gil-Av, Ph.D.
Professor
Department of Organic Chemistry
Weizmann Institute of Science
Rehovot, Israel

Petr Hušek, Ph.D.
Senior Scientist
Research Institute of Endocrinology
Prague, Czechoslovakia

Floyd E. Kaiser, M.S.
Supervisor of Pharmaceutical
Analysis and Chemist IV
Analytical Biochemistry Laboratories,
 Inc.
Columbia, Missouri

Wilfried A. König, Ph.D.
Institut für Organische Chemie
Universität Hamburg
Hamburg, West Germany

Kenneth C. Kuo, M.S.
Senior Research Chemist-
  Chromatographer
Department of Biochemistry and
  Experiment Station Chemical Labs
University of Missouri-Columbia
Columbia, Missouri

Shigefumi Kuramoto, M.S.
Japan Catalytic Chemical
  Industrial Cooperation
Osaka, Japan

Samuel L. MacKenzie, Ph.D.
Senior Research Officer
Plant Biotechnology Institute
National Research Council of Canada
Saskatoon, Canada

Masami Makita, Ph.D.
Professor of Health Chemistry
Faculty of Pharmaceutical Sciences
Okayama University
Okayama, Japan

Iain MacArthur Moodie, Ph.D.
Specialist Scientist
South African Medical Research Council
Metabolic Research Group
Tygerberg Hospital
Cape Town, South Africa

Graeme J. Nicholson, Ph.D.
Institut für Organische Chemie
Universität Tübingen
Tübingen, West Germany

Göran Odham
Associate Professor and Director
Laboratory of Ecological Chemistry
Ecology Building
University of Lund
Lund, Sweden

Naobumi Ôi
Managing Director
Sumika Chemical Analysis Service, Ltd.
Osaka, Japan

Prudent Padieu
Professor
Department of Medical Biochemistry
Faculté de Médecine
Centre Hospitalier Régional Universitaire
Université de Dijon
Dijon, France

Cyril Ponnamperuma, Ph.D.
Professor of Chemistry
Director of Laboratory of Chemical
  Evolution
Department of Chemistry
University of Maryland
College Park, Maryland

James E. Pautz
Department of Biochemistry and
  Experiment Station Chemical Labs
University of Columbia-Missouri
Columbia, Missouri

Larry L. Wall, Sr., M.S.
Process Engineering Supervisor
3M Company
Columbia, Missouri

Shigeo Yamamoto, Ph.D.
Associate Professor of Health Chemistry
Faculty of Pharmaceutical Sciences
Okayama University
Okayama, Japan

Robert W. Zumwalt, Ph.D.
Research Associate
Department of Biochemistry
University of Missouri-Columbia
Columbia, Missouri

TABLE OF CONTENTS

Volume I

Volume II

Volume III

Chapter 1

ANALYSIS OF AMINO ACIDS AND *N*-METHYLAMINO ACIDS BY CAPILLARY CHROMATOGRAPHY OF DIASTEREOMERIC DERIVATIVES: THE USE OF ISOCYANATES AS REAGENTS FOR ENANTIOMER SEPARATION

Wilfried A. König

TABLE OF CONTENTS

I. INTRODUCTION

Configurational studies of amino acids require highly efficient separation techniques. Capillary gas chromatography (GC) exhibits a unique combination of substrate specificity and sensitivity and is therefore particularly suited for the separation of stereoisomers.

Stereoisomers can either be separated as diastereomers or as enantiomers, depending on the selectivity of the GC column. Diastereomeric derivatives are obtained by reacting amino acids with an appropriate chiral reagent and may be separated on any temperature-stable stationary phase. These separations could be considerably improved by the development of modern glass and fused silica capillary GC in recent years. Capillary columns with a large variety of stationary phases are commercially available today.

For direct enantiomer separation chiral stationary phases are required. The separation of amino acid enantiomers is brought about by enantioselective molecular interactions with the chiral molecules of the stationary phase. Hydrogen bonding, dipole-dipole interactions, and electronic interactions may contribute to the formation of diastereomeric association complexes with different amounts of internal energy. The enantioselectivity of such a GC system strongly depends on the structural properties of the chiral stationary phase and on the type of derivative used. New methods in the preparation of temperature-stable chiral phases and in the formation of new types of derivatives have greatly improved the techniques of enantiomer separation, not only for amino acids but for various other compound classes as well.[1]

II. CONFIGURATIONAL INVESTIGATIONS OF AMINO ACIDS — PRACTICAL APPLICATIONS

In spite of the prevailing occurrence of L-amino acids in nature, there is a demand in microtechniques for the determination of the stereochemical properties of amino acids. D-Enantiomers regularly occur as constituents of peptide antibiotics. These natural products are also a source of unusual and even new amino acids with unknown configuration.

From the first detection of a D-amino acid in a vertebrate — D-alanine as a constituent of dermorphin,[2] a peptide isolated from frog skin — the necessity of configurational analysis of natural peptides gained even more importance.

The scope of applications for GC as a technique for amino acid configurational analysis also includes the investigation and proof of retention of configuration during the different steps of peptide synthesis,[3,4] the proof of optical purity of amino acids as pharmaceuticals, and the determination of the optical yield during asymmetric synthesis of amino acids.

III. THE USE OF DIASTEREOMERIC DERIVATIVES

The separation of diastereoisomers by GC has found broad application in configurational analysis of amino acids and many other compounds.[5] Diastereomers are obtained by the introduction of a second asymmetric center into the amino acid molecule. This may be achieved by esterificiation of the carboxylic group with chiral alcohols, by formation of an amide with a chiral amine or amino acid ester, or by acylation of the amino group with a chiral acylation reagent.

A few general remarks have to be made concerning the reliability and precision of stereochemical investigations with diastereomeric derivatives. The reliability of configurational assignment usually depends on empirical findings. So the order of elution of enantiomers is D before L in the case of an amino acid esterified with (*S*)-(+)-3-methyl-2-butanol and acylated with trifluoroacetyl (TFA) or pentafluoropropionyl (PFP) groups. The same order of elution is observed when using *N*-trimethylsilyl (TMS) derivatives.

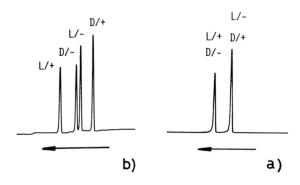

FIGURE 1. (a) Separation of D,L-valine-($\pm$)-3-methyl-2-bu-tyl-esters; 25-m fused silica capillary with achiral CpSil-5 (Chrompack®), 120°C; carrier gas, 0.6 bar H_2. (b) Separation of same sample on a 40-m glass capillary with chiral XE-60-L-valine-(S)-α-phenylethyl amide, 140°C; carrier gas, 0.8 bar H_2.

The precision of determination of the proportion of a mixture of enantiomers or of the optical purity mainly depends on the optical purity of the chiral reagent, which is never 100%. If small amounts of D-enantiomers are to be determined in a test for racemization of L-amino acids and (S)-(+)-3-methyl-2-butanol is used as a chiral reagent, an impurity of (R)-(−)-3-methyl-2-butanol forms a derivative with the L-enantiomer, which coelutes with the (S)-(+)-ester of the D-enantiomer (the L/(S)-(+)- and D/(R)-(−)-derivatives are enantiomers). This means that the L/(R)-(−)-peak adds to the area of the D/(S)-(+)-peak and a serious error may be introduced if the proportion of the optical impurity of the alcohol is not known. This problem can be avoided by using chiral stationary phases. On these phases the enantiomeric contribution introduced by the optical impurity of the reagent can be separated and the error due to the optical impurity can be eliminated[6] (Figure 1).

A further uncertainty may arise from different reaction kinetics of enantiomers in the formation of diastereomeric derivatives. The diastereomeric transition states in the reaction of two chiral compounds have different energy contents. Consequently, the reaction of a chiral reagent with *one* enantiomer may proceed faster than with its antipode. Particularly in cases of reactions with poor yields the thermodynamic equilibrium is not reached and a considerable error has to be accounted for by this kinetic resolution.

A. Chiral Reagents for the Formation of Diastereoisomers

There are two routes for obtaining diastereomeric derivatives of amino acids. A common procedure is the esterification with a chiral alcohol. 2-Butanol,[7,8] 3-methyl-2-butanol (Structure 1),[9-12] 3,3-dimethyl-2-butanol,[12,13] 2-octanol,[8] and menthol[14,15] have been the most commonly used reagents.

$$\begin{matrix} H_3C \\ H_3C \end{matrix} \diagdown CH-CH-CH_3 \\ \qquad \underset{OH}{|}$$

<u>1</u>

Alternatively, the amino function may be used for conversion of an amino acid into a diastereomeric derivative. Usually chlorides of chiral acids such as N-TFA-L-prolyl chloride,[16] L-α-chloroisovaleryl chloride (Structure 2),[17,18] or chloroformates of chiral alcohols[19] are used.

$$\text{H}_3\text{C} \diagdown \text{CH-CH-C} \diagup \text{O} \atop \text{H}_3\text{C} \diagup \quad \text{Cl} \quad \text{Cl}$$

2

The remaining functions of the amino acid have to be derivatized by esterification or acylation, respectively. Side chain functions may be trimethylsilylated, if necessary, e.g., in the case of hydroxy groups.

Some of these procedures have been optimized by applying glass capillary GC. The use of (+)-3-methyl-2-butanol and L-α-chloroisovaleryl chloride as chiral reagents will be discussed in detail in this chapter.

B. *N*-Acylated Amino Acid-(+)-3-Methyl-2-Butyl Esters

1. Preparation of S-(+)-3-Methyl-2-Butanol (Structure 1)

A convenient method of preparing (+)-3-methyl-2-butanol was described by Halpern and Westley.[20] L-Valine is esterified with racemic 3-methyl-2-butanol under catalytic action of *p*-toluene sulfonic acid and under continuous removal of the water formed during the reaction. After two recrystallizations of the toluene sulfonic acid salt of L-valine-(±)-3-methyl-2-butyl ester the ratio of (+)-3-methyl-2-butyl ester to (−)-3-methyl-2-butyl ester is 98.5 to 1.5% (as determined by GC and peak integration). Subsequent hydrolysis of the L-valine-(+)-ester with NaOH, extraction with diethyl ether, and distillation yields (+)-3-methyl-2-butanol in an overall yield of 34%.[21]

2. Formation of Derivatives[10]

For esterification, samples of 100 μg of amino acid were heated for 90 min at 100°C in 150 μℓ of a 7 *M* solution of dry HCl gas in (+)-3-methyl-2-butanol in a screw cap vial with Teflon® lining in the cap. For histidine, ornithine, lysine, and arginine the methyl esters were formed in the initial step by heating the samples in a 1.5 *M* solution of HCl gas in methanol for 1 hr at 100°C. After removal of excess reagent the methyl esters were transesterified with (+)-3-methyl-2-butanol as described above. The reagent was then removed by a stream of nitrogen, and 200 μℓ of dichloromethane and 50 μℓ of PFP anhydride were added. After 30 min at room temperature the reagent was removed with nitrogen and the sample was dissolved in ethyl acetate for GC. Instead of the PFP derivatives the corresponding TFA derivatives can be used. However, the PFP derivatives are more volatile and show less peak overlapping in GC separation.

3. Capillary GC

By using this derivatization procedure the stereoisomers of all amino acids commonly found in proteins could be separated, including histidine, arginine, and tryptophan (Table 1).[10] For GC a 25-m glass capillary coated with SE 30 (LKB, Type 2101-210, AmAc) was used. In all cases the derivatives of the L-enantiomers are eluted after the D-enantiomer (Figures 2 and 3). Since the basic amino acids histidine, arginine, and tryptophan are incompletely acylated at room temperature, a small amount of PFP anhydride was injected together with the sample, as already described by Gehrke and Stalling.[22]

4. Configurational Investigations of Unusual Amino Acids

Although stereochemical investigations of natural amino acids have often successfully been performed by enantiomer separation on chiral stationary phases,[23-25] there were several cases when this technique failed. The amino acids (Structure 3) found as constituents of the nikkomycins,[26,27] a group of peptide-nucleoside antibiotics, could not be eluted from chiral phases so far. The L-configuration at the α-carbon atom was assigned to these amino acids according to the results obtained with (+)-3-methyl-2-butyl esters,[28] although no racemic reference compound was available.

Table 1
SEPARATION FACTORS (α) AND
OPERATING TEMPERATURES FOR
SEPARATION OF *N,O*-PFP-D,L-AMINO
ACID-(+)-3-METHYL-2-BUTYL
ESTERS AND *N*-METHYL-D,L-AMINO
ACID-(+)-3-METHYL-2-BUTYL
ESTERS

Amino acid	α	Column temp. (°C)	Column
Ala	1.09	100	a
Abu[1]	1.09	100	a
Thr	1.03	100	a
Val	1.08	100	a
Ser	1.05	100	a
Leu	1.09	100	a
*a*Ile	1.08	100	a
Ile	1.09	100	a
Cys	1.07	100	a
Pro	1.06	100	a
Dap[2]	1.04	140	a
Met	1.05	140	a
Orn	1.06	140	a
Phe	1.04	140	a
Asp	1.04	140	a
His	1.03	140	a
Lys	1.05	140	a
Tyr	1.04	140	a
Glu	1.05	140	a
Arg	1.05	140	a
Trp	1.04	200	a
N-Me-Ala	1.038	80	b
N-Me-Abu	1.041	80	b
N-Me-Val	1.030	100	b
N-Me-Leu	1.024	100	b
N-Me-Ile	1.032	100	b
N-Me-*a*Ile	1.026	100	b

[1]Abu = α-amino-*n*-butyric acid; [2]Dap = 2,3-diamino-propionic acid; a = 25 m glass capillary column, SE 30 (AmAc, LKB 2101-210); b = 25 m glass capillary column, CpSil-5 (Chrompack®).

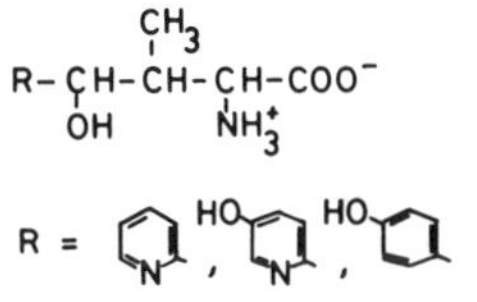

$$R-\underset{\underset{OH}{|}}{C}H-\underset{\underset{CH_3}{|}}{C}H-\underset{\underset{NH_3^+}{|}}{C}H-COO^-$$

R =

3

a. Determination of Configuration

There is one important advantage in using diastereomeric derivatives instead of direct enantiomer separation on chiral stationary phases for determination of a configuration.

Usually, natural amino acids occur in only *one* enantiomeric form (L and sometimes D),

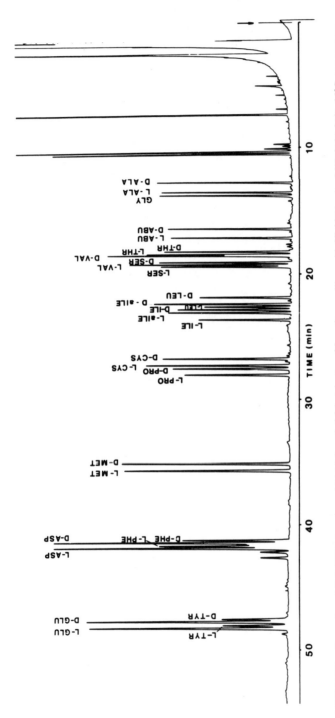

FIGURE 2.   Separation of *N,O*-PFP-D,L-amino acid-(+)-3-methyl-2-butyl esters; 25-m glass capillary with SE-30 (AmAc, LKB 2101-210), 85°C, temperature program, 2°/min to 220°C; carrier gas, 0.7 bar H₂.

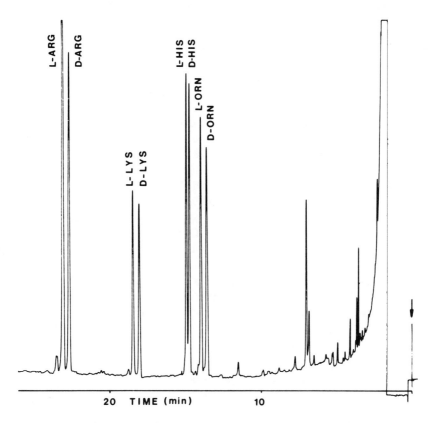

FIGURE 3. Separation of *N*-PFP derivatives of basic D,L-amino acid-(+)-3-methyl-2-butyl esters. Column as in Figure 2, 150°C, temperature program, 2°/min to 220°C.

therefore give only *one* peak on a chiral column when: (1) the amino acid is derivatized without the introduction of a second asymmetric center, or (2) when the amino acid is derivatized with one isomer of a chiral alcohol as shown in Scheme 1b and 1c.

SCHEME 1

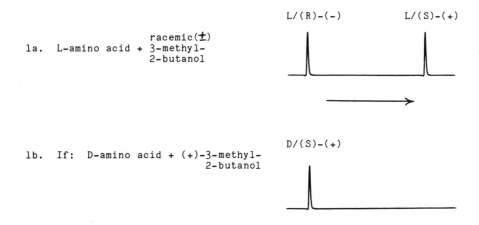

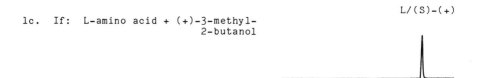

1c.   If: L-amino acid + (+)-3-methyl-
                          2-butanol

Thus, with these types of derivatization the configuration cannot be derived if the racemate of this amino acid is not available for comparison and coinjection. However, if this amino acid is derivatized with racemic (±)-3-methyl-2-butanol, two peaks are obtained in the chromatogram (Scheme 1a). For example, if the amino acid has the L-configuration, the L/(R)-(−)- and L/(S)-(+)-diastereomers are formed. The L/(S)-(+)-isomer is always eluted after the L/(R)-(−)-isomer. By using (+)-3-methyl-2-butanol as chiral reagent it is now very easy to find out if a D- or L-amino acid is present in the sample. The D/(S)-(+)-diastereomer would be eluted together with the first peak, the L/(R)-(−)-isomer (Scheme 1b), and the L/(S)-(+)-isomer would coelute with the second peak when the mixture is injected (Scheme 1c).

The configuration of isovaline (Structure 4), a constituent of the peptide antibiotic trichotoxin A-40, could also be determined by this method.[29] It has to be noted that, in this case of an α-alkylated amino acid, the L/(S)-(+)-isomer is eluted before the D/(S)-(+)-isomer.

$$H_3C-CH_2-\underset{\underset{NH_3^+}{|}}{\overset{\overset{CH_3}{|}}{C}}-COO^-$$

4

5. N-Methylamino Acids

This group of modified amino acids is commonly found in many peptide antibiotics. When *N*-methylamino acid esters are submitted to acylation in the usual way with TFA anhydride the expected derivatives are obtained in very poor yields or not at all. Instead of *N*-acyl-*N*-methylamino acid esters, heterocyclic products, substituted alkylidene-oxazolidine-5-ones are formed under complete racemization (Figure 4).[30] The (+)-3-methyl-2-butyl esters of *N*-methylamino acids can be separated without further derivatization, but badly tailing peaks are obtained.

It was found, however, that symmetrical peak form is observed when a small amount of a silylation reagent such as *N*-methyl-*N*-trimethylsilyl-trifluoroacetamide (MSTFA) is co-injected with the sample (Figure 5).[31] The silylation reagent temporarily deactivates the inner surface of the capillary column. In all cases the derivatives of the L-enantiomers have longer retention times than the D-enantiomers (Figure 6). Alternatively, the *N*-methyl-*N*-TMS amino acid-(+)-3-methyl-2-butyl esters can be formed, but in some cases the separation of the diastereomers is not complete. *N*-Trimethylsilylation reverses the order of elution, with the (+)-3-methyl-2-butyl esters of the L-enantiomers eluted before the D-form.

6. Other α-Substituted Acids

(+)-3-Methyl-2-butanol can also be applied to stereochemical investigations of α-hydroxy acids and α-branched carboxylic acids.[32] The hydroxy groups of the diastereomeric esters are either trimethylsilylated or trifluoroacetylated. As in the case of amino acids, the derivatives of L-hydroxy acids have longer retention times than the derivatives of the D-enantiomers. For α-alkylated acids the order of elution has not yet been determined.

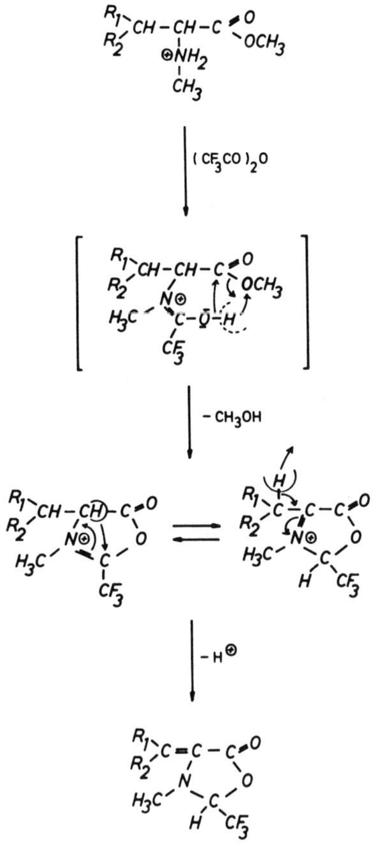

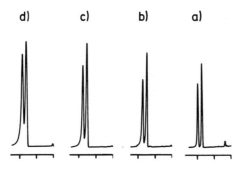

FIGURE 4.   Reaction of *N*-methylamino
acid methyl esters with TFA anhydride.

FIGURE 5.   Separation of *N*-methyl-D,L-valine-(+)-
3-methyl-2-butyl ester. (a) Co-injection of 0.5 μℓ
of MSTFA; (b) reinjection without MSTFA after 10
min; (c) after 20 min; (d) after 40 min. Column as
in Figure 1a).

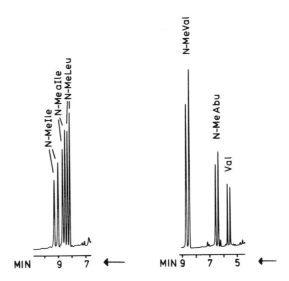

FIGURE 6.   Separation of (+)-3-methyl-2-butyl esters of (right)
N-methyl-D,L-valine, *N*-methyl-D,L-α-aminobutyric acid (90°C)
and (left) *N*-methyl-D,L-leucine, *N*-methyl-D,L-*allo*-isoleucine,
and *N*-methyl-D,L-isoleucine (100°C). Column as in Figure 1a,
co-injection of 0.5 μℓ of MSTFA.

C. L-α-Chloroisovaleryl Derivatives

L-α-Chloroisovaleryl chloride was first proposed as a derivatizing reagent by Halpern and
Westley.[17] The reagent can be prepared from L-valine by deamination with sodium nitrite
in 6 *N* HCl and subsequent substitution on the α-carbon atom by chlorine according to
Greenstein et al.[33] This reaction proceeds with retention of configuration due to neighboring
group participation of the carboxylic function. In the last step the acid chloride is formed
by reaction with thionyl chloride.

The reagent is used for acylation of amino acid methyl esters as well as for chiral amines
and amino alcohols at room temperature.[18] In the case of amino acid esters the derivatives
of the L-enantiomers elute after the derivatives of the D-enantiomers. The results of GC
investigations with capillary columns with unpolar and chiral stationary phases demonstrate
that this method may be very useful (Figure 7).

IV. SEPARATION OF AMINO ACID ENANTIOMERS ON CHIRAL STATIONARY PHASES

After the first successful enantiomer separation by Gil-Av and associates,[34,35] much effort
was undertaken to improve the thermostability and enantioselectivity of chiral stationary
phases.[36-38] A great step in this direction was achieved by Bayer and co-workers[39,40] in
preparing chiral polysiloxanes by copolymerization of dimethylsiloxane with carboxyalkyl-
methyl-siloxanes and incorporation of L-valine-*tert*-butylamide (Chirasil-Val®). Already,
Feibush[41] has shown with *N*-dodecanoyl-L-valine-*tert*-butylamide that this residue has high
enantioselectivity. Similar chiral stationary phases can be prepared by modification of
polysiloxanes with cyanoalkyl side chains as demonstrated by Verzele et al. with OV-225[42]
and Silar®-10C[43] and by König et al. with XE-60[44] and other polysiloxanes (Figure 8).
Particularly XE-60-L-valine-(*S*)-α-phenylethylamide and the corresponding (*R*)-α-phenyl-
ethylamide were shown to have high thermostability and enantioselectivity towards amino
acids,[44] amino alcohols,[45] and carbohydrates.[45-47]

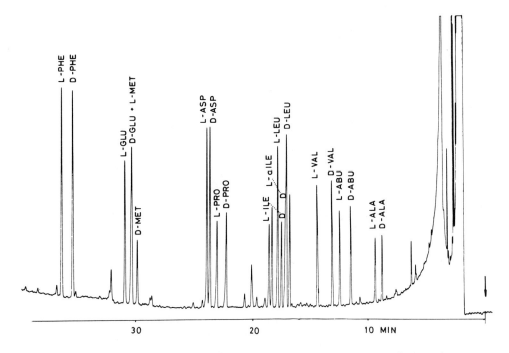

FIGURE 7.   Separation of L-α-chloroisovaleryl-D,L-amino acid methyl esters. Column as in Figure 2; 150°C, 9 min isothermal, then temperature program 2°/min to 230°C.

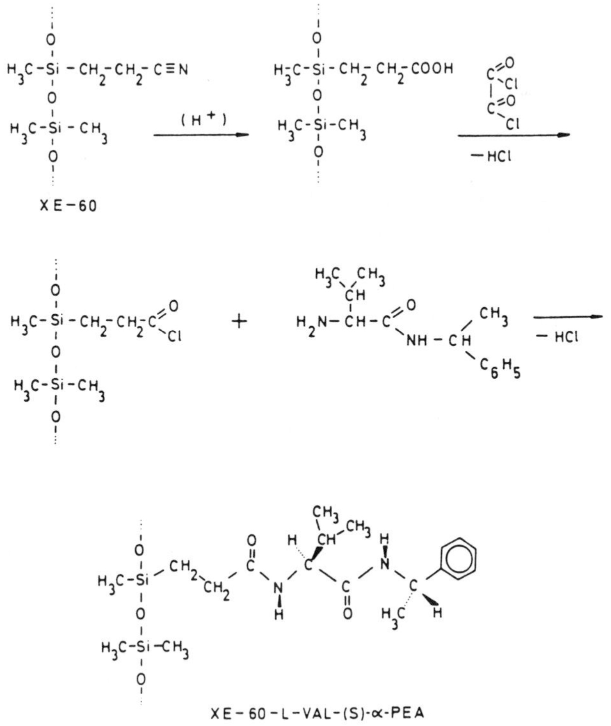

FIGURE 8.   Reaction scheme for the preparation of XE-60-L-valine-(S) or (R)-α-phenylethyl amide.

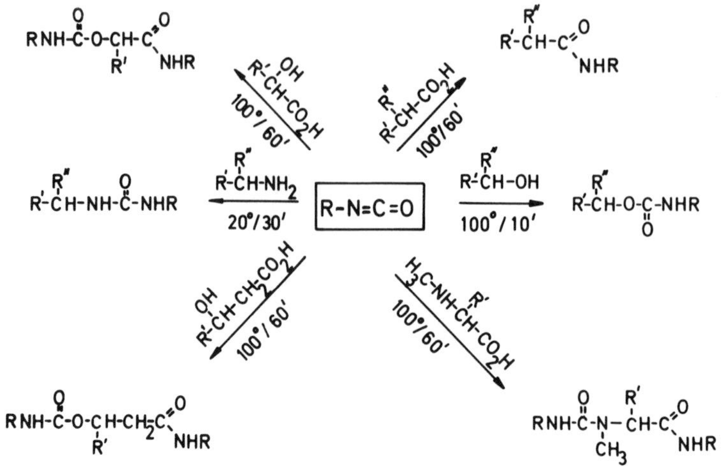

FIGURE 9.   Reaction of isocyanates with compounds having hydroxy-, amino-, *N*-methylamino-, and carboxylic groups.

A. Isocyanates as Universal Reagents for Enantiomer Separation

GC enantiomer separation is achieved by diastereomeric association of chiral substrates with a chiral stationary phase. This molecular interaction is strongly influenced by the type and number of interacting functional groups of both the stationary phase *and* the substrate. In the past much attention was paid to structural modification of known chiral stationary phases, using always the same highly volatile derivatives of the substrates. With the advent of thermostable stationary phases, less volatile derivatives can be used. We have demonstrated that chiral alcohols[48] and α-hydroxy acid esters[49] after reaction with isopropyl isocyanate can be separated as *N*-isopropyl carbamates.[50] The same reagent can be used to convert chiral amines into *N*-isopropyl ureido derivatives.[49] They are also separated with high separation factors. The versatility of isopropyl isocyanate and other isocyanates (Figure 9) was further demonstrated in the reaction with free carboxylic groups, which are converted into amides. Although introduction of a further polar amide group decreases the volatility of the derivatives, the enantioselectivity is considerably enhanced and it is now possible for the first time to separate the enantiomers of *N*-methylamino acids (Figure 10) and β-hydroxy acids as *N*-isopropyl carbamates/*N*-isopropyl amides or *N*-isopropyl ureido derivatives/*N*-isopropyl amides, respectively.[51]

It could be proved that the formation of these derivatives proceeds without racemization in the case of *N*-methylamino acids and α- and β-hydroxy acids.

N-Alkyl-ureido derivatives can also be prepared with amino acid esters. Particularly in cases when *N*-acylated amino acid esters are difficult to separate, as in the case of proline or isovaline, the formation of *N*-ureido derivatives may be favorable.

By heating the amino acid isopropyl esters in a mixture of 100 μℓ of dichloromethane and 100 μℓ of isopropyl isocyanate together with 5 μℓ of pyridine, mixtures of mono- and bis-substituted *N*-isopropyl-ureido derivatives are usually formed, while in the case of *tert* butyl isocyanate mono-ureido derivatives are formed exclusively, as proven by GC/MS investigations.[52] The reaction also proceeds without racemization. A separation of the *N*-tert-butyl-ureido derivatives of some aliphatic amino acid isopropyl esters and of proline is shown in Figure 11. The separation factors for some additional amino acids are given in Table 2.

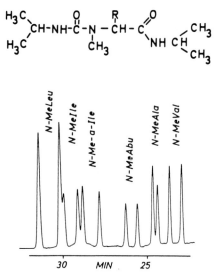

FIGURE 10. Separation of *N*-methyl-D,L-amino acids as *N*-isopropyl-ureido/*N*-isopropyl amide derivatives; 25-m glass capillary with XE-60-L-valine-(*R*)-α-phenylethyl amide; 170°C, carrier gas, 0.6 bar H₂. L-Enantiomers are eluted first.

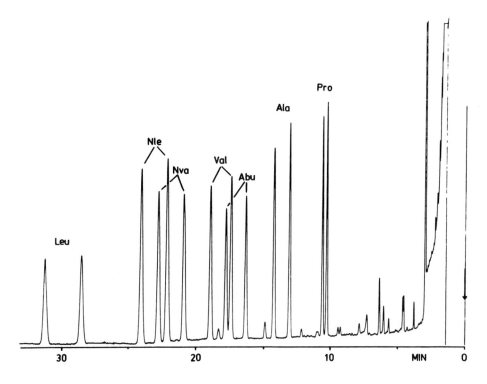

FIGURE 11. Separation of *N-tert*-butyl-ureido-D,L-amino acid-isopropyl esters; 25-m fused silica column with XE-60-L-valine-(*S*)-α-phenylethyl amide (Chrompack®); 180°C; carrier gas, 0.6 bar H₂. D-Enantiomers are eluted first. Abu = α-amino-*n*-butyric acid; Nva = norvaline; Nle = norleucine.

Table 2
SEPARATION FACTORS (α)
AND OPERATING
TEMPERATURES FOR
SEPARATION OF *N-TERT-*
BUTYL-UREIDO-D,L-AMINO
ACID-ISOPROPYL ESTERS ON
A 15 M GLASS CAPILLARY
COLUMN WITH XE-60-L-
VALINE-(S-α-PHENYLETHYL
AMIDE (IN ORDER OF
EMERGENCE)

Amino acid	α	Column temp.(°C)
Iva[a]	1.02	170
Pro	1.03	180
Ala	1.09	180
Abu	1.08	180
Val	1.08	180
Nval	1.09	180
aIle	1.09	180
Thr (OTMS)[b]	1.09	180
aThr (OTMS)[b]	1.05	180
Leu	1.09	180
Ile	1.08	180
Ser (OTMS)[b]	1.11	180
Nle	1.09	180
Asp	1.08	180
Met	1.07	200
Phg[c]	1.02	200
Glu	1.07	200
Phe	1.07	200

[a]   Iva = isovaline.
[b]   Side chain hydroxy group was silylated with
      MSTFA.
[c]   Phenylglycine.

V. CONCLUDING REMARKS

Modern GC with glass and fused silica capillary columns has stimulated the development of highly efficient separation techniques for stereoisomers either as diastereomeric derivatives or by direct enantiomer separation on chiral stationary phases. New thermostable polysiloxanes with incorporated chiral molecules and new types of derivatives, supporting diastereomeric interaction between chiral stationary phase and substrate enable the analyst to solve stereochemical problems in the field of amino acids and related compounds. These techniques proved to be valuable tools for configurational analysis of natural and synthetic peptides, of amino acids and peptides with pharmacological importance, and help to control stereoselectivity of asymmetric syntheses and biotechnological processes.

ACKNOWLEDGMENTS

I have to thank all my co-workers who have contributed to the results of my research group cited in this article, particularly Dr. I. Benecke, Dr. K. Kruse, N. Lucht, Dr. W. Rahn, E. Schmidt, J. Schulze, Dr. S. Sievers, and Dr. K. Stölting. The financial support of our work by *Deutsche Forschungsgemeinschaft* and *Fonds der Chemischen Industrie* is gratefully acknowledged.

Gratitude is also expressed to Elsevier Scientific Publishing for granting permission to use figures from the *Journal of Chromatography*.

REFERENCES

1. **König, W. A.,** Separation of enantiomers by capillary gas chromatography with chiral stationary phases, *J. High Resolut. Chromatogr. Chromatogr. Commun.,* 5, 588, 1982.
2. **de Castilione, R., Faoro, F., Perseo, G., and Piani, S.,** *Proc. 16th Eur. Peptide Symp.,* Helsingor, Brunfeldt, K., Ed., Scriptor, Copenhagen, 1981, 441.
3. **Weygand, F., Prox, A., Schmidhammer, L., and König, W.,** Gas chromatographic investigation of racemization during the synthesis of peptides, *Angew. Chem.* 75, 282, 1963; *Angew. Chem. Int. Ed. Engl.,* 2, 183, 1963.
4. **Bayer, E., Gil-Av, E., König, W. A., Nakaparksin, S., Oro, J., and Parr, W.,** Retention of configuration in the solid phase synthesis of peptides, *J. Am. Chem. Soc.,* 92, 1738, 1970.
5. **Halpern, B.,** Derivatives for chromatographic resolution of optically active compounds, in *Handbook of Derivatives for Chromatography,* Blau, K. and King, G., Eds., Heyden & Son, London, 1977, 457.
6. **König, W. A.,** Gas-chromatographische trennung von diastereomeren aminosaureestern an chiralen stationaren phasen, *Chromatographia,* 9, 72, 1976.
7. **Pollock, G. E., Oyama, V. I., and Johnson, R. D.,** Resolution of racemic amino acids by gas chromatography, *J. Gas Chromatogr.,* 3, 174, 1965.
8. **Gil-Av, E., Charles-Sigler, R., and Fischer, G.,** Resolution of amino acids by gas chromatography, *J. Chromatogr.,* 17, 408, 1965.
9. **Westley, J. W., Halpern, B., and Karger, B. L.,** Effect of solute structure on separation of diastereoisomeric esters and amides by gas-liquid chromatography, *Anal. Chem.,* 40, 2046, 1968.
10. **König, W. A., Rahn, W., and Eyem, J.,** Gas chromatographic separation of diastereoisomeric amino acid derivatives on glass capillaries, *J. Chromatogr.,* 133, 141, 1977.
11. **Rahn, W., Eckstein, H., and König, W. A.,** A micro method for the determination of the configuration of histidine in peptides: evidence for partial racemization during peptide synthesis, *Hoppe-Seyler's Z. Physiol. Chem.,* 357, 1223, 1976.
12. **Ayers, G. S., Monroe, R. E., and Mossholder, J. H.,** Resolution of amino acid diastereomers by means of packed column gas chromatography, *J. Chromatogr.,* 63, 259, 1971.
13. **Pollock, G. E. and Kawauchi, A. H.,** Resolution of racemic aspartic acid, tryptophan, hydroxy and sulfhydryl amino acids by gas chromatography, *Anal. Chem.,* 40, 1356, 1968.
14. **Halpern, B. and Westley, J. W.,** Resolution of neutral DL-amino acids via their (−)-menthyl ester derivatives, *J. Chem. Soc. Chem. Commun.,* 421, 1965.
15. **Hasegawa, M. and Matsubara, I.,** Gas chromatographic determination of the optical purities of amino acids using N-trifluoroacetyl menthyl esters, *Anal. Biochem.,* 63, 308, 1975.
16. **Halpern, B. and Westley, J. W.,** High sensitivity optical resolution of D,L amino acids, *Biochem. Biophys. Res. Commun.,* 19, 361, 1965.
17. **Halpern, B. and Westley, J. W.,** High sensitivity optical resolution of DL-amino acids by gas chromatography, *J. Chem. Soc. Chem. Commun.,* 246, 1965.
18. **König, W. A., Stoelting, K., and Kruse, K.,** Gas chromatographic separation of optically active compounds in glass capillaries, *Chromatographia,* 10, 444, 1977.
19. **Halpern, B. and Westley, J. W.,** The use of (−)-menthyl chloroformate in the optical analysis of asymmetric amino and hydroxy compounds by gas chromatography, *J. Org. Chem.,* 33, 3978, 1968.
20. **Halpern, B. and Westley, J. W.,** Chemical resolution of secondary (+)-alcohols, *Aust. J. Chem.,* 19, 1533, 1966.
21. **Benecke, I.,** Diplom Thesis, Universitat Hamburg, 1980.
22. **Gehrke, C. W. and Stalling, D. L.,** Quantitative analysis of the twenty natural protein amino acids by gas-liquid chromatography, *Separ. Sci.,* 2, 101, 1967.
23. **König, W. A., Kneifel, H., Bayer, E., Müller, G., and Zähner, H.,** Metabolic products of micro organisms 116* O-{L-Norvalyl-5}-isourea, a new arginine antagonist, *J. Antibiot.,* 26, 44, 1973.
24. **König, W. A., Loeffler, W., Meyer, W. H., and Uhmann, R.,** L-Arginyl-D-*allo*-threonyl-L-phenylalanin, ein aminosäure-antagonist aus dem pilz keratinophyton terreum, *Chem. Ber.,* 106, 816, 1973.

25. **König, W. A., Engelfried, C., Hagenmaier, H., and Kneifel, H.,** Struktur des peptidantibiotikums stenothricin, *Liebigs Ann. Chem.,* 2011, 1976.

26. **Fiedler, E., Fiedler, H. P., Gerhard, A., Keller-Schierlein, B., König, W. A., and Zähner, H.,** Stoffwechselprodukte von mikroorganismen, *Arch. Microbiol.,* 107, 249, 1976.

27. **König, W. A., Hass, W., Dehler, W., Fiedler, H.-P., and Zähner, H.,** Strukturafklarung und partialsynthese des nucleosidantibiotikums nikkomycin B, *Liebigs Ann. Chem.,* 622, 1980.

28. **König, W. A., Pfaff, K.-P., Bartsch, H.-H., Schmalle, H., and Hagenmaier, H.,** Konfiguration von 2-Amino-4-hydroxy-4-(5-hydroxy-2-pyridyl)-3-methylbuttersäure, der N-terminalen aminosäure der nikkomycine, *Liebigs Ann. Chem.,* 1728, 1980.

29. **Brückner, H., Nicholson, G. J., Jung, G., Kruse, K., and König, W. A.,** Gas chromatographic determination of the configuration of isovaline in antiamoebin, samarosporin, (emerimicin IV), stilbellin, suzukacillins and trichotoxins, *Chromatographia,* 13, 209, 1980.

30. **König, W. A. and Hess, U.,** Reaktionen von Aminosäuren und Peptiden, I: Die Reaktion von N-Methylvalin und N-Methylisoleucin mit Trifluoracetanhydrid, *Liebigs Ann. Chem.,* 1087, 1977.

31. **König, W. A., Benecke, I., and Schulze, J.,** Configurational analysis and test of racemization of *N*-methylamino acids by capillary gas chromatography, *J. Chromatogr.,* 238, 237, 1982.

32. **König, W. A. and Benecke, I.,** Gas chromatographic separation of chiral 2-hydroxy acids and 2-alkyl-substituted carboxylic acids, *J. Chromatogr.,* 195, 292, 1980.

33. **Fu, S. C. J., Birnbaum, S. M., and Greenstein, J. P.,** Influence of optically active acyl groups on the enzymatic hydrolysis of *N*-acylated-L-amino acids, *J. Am. Chem. Soc.,* 76, 6054, 1954.

34. **Gil-Av, E., Feibush, B., and Charles-Sigler, R.,** in *Gas Chromatography 1966,* Littlewood, A. B., Ed., Institute of Petroleum, London, 1967, 227.

35. **Gil-Av, E. and Feibush, B.,** Resolution of enantiomers by gas-liquid chromatography with optically active stationary phases. Separation on packed columns, *Tetrahedron Let.,* 3345, 1967.

36. **König, W. A., Parr, W., Lichtenstein, H. A., Bayer, E., and Oro, J.,** Gas chromatographic separation of amino acids and their enantiomers: nonpolar stationary phases and a new optically active phase, *J. Chromatogr. Sci.,* 8, 183, 1970.

37. **König, W. A. and Nicholson, G. J.,** Glass capillaries for fast gas chromatographic separation of amino acid enantiomers, *Anal. Chem.,* 47, 951, 1975.

38. **Charles, R., Beitler, U., Feibush, B., and Gil-Av, E.,** Separation of enantiomers on packed columns containing optically active diamide phases, *J. Chromatogr.,* 112, 121, 1975.

39. **Frank, H., Nicholson, G. J., and Bayer, E.,** Chiral polysiloxanes for resolution of optical antipodes, *Angew. Chem.,* 90, 396, 1978; *Angew. Chem. Int. Ed. Engl.,* 17, 363, 1978.

40. **Frank, H., Nicholson, G. J., and Bayer, E.,** Enantiomer labelling: a method for the quantitative analysis of amino acids, *J. Chromatogr.,* 167, 187, 1978.

41. **Feibush, B.,** Interaction between asymmetric solutes and solvents, *J. Chem. Soc., Chem. Commun.,* 545, 1971.

42. **Saeed, T., Sandra, P., and Verzele, M.,** Synthesis and properties of a novel chiral stationary phase for the resolution of amino acid enantiomers, *J. Chromatogr.,* 186, 611, 1979.

43. **Saeed, T., Sandra, P., and Verzele, M.,** GC separation of the enantiomers of proline and secondary amines, *J. High Resolut. Chromatogr. Chromatogr. Commun.,* 3, 35, 1980.

44. **König, W. A., Sievers, S., and Benecke, I.,** New optically active stationary phases for enantiomer separation, in *Proc. 4th Int. Symp. Capillary Chromatogr.,* Dr. A. Hüthig, Heidelberg, 1981, 703.

45. **König, W. A., Benecke, I., and Sievers, S.,** New results in the gas chromatographic separation of enantiomers of hydroxy acids and carbohydrates, *J. Chromatogr.,* 217, 71, 1981.

46. **König, W. A., Benecke, I., and Bretting, H.,** Gas-chromatographische trennung enantiomerer kohlenhydrate an einer neuen chiralen stationären phase, *Angew. Chem.,* 93, 668, 1981; *Angew. Chem. Int. Ed. Engl.,* 20, 693, 1981.

47. **Benecke, I., Schmidt, E., and König, W. A.,** Gas chromatographic resolution of carbohydrate enantiomers. A new chiral phase for pentoses, *J. High Resolut. Chromatogr. Chromatogr. Commun.,* 4, 553, 1981.

48. **König, W. A., Francke, W., and Benecke, I.,** Gas chromatographic enantiomer separation of chiral alcohols, *J. Chromatogr.,* 239, 227, 1982.

49. **König, W. A., Benecke, I., and Sievers, S.,** New procedure for gas chromatographic enantiomer separation. Application to chiral amines and hydroxy acids, *J. Chromatogr.,* 238, 427, 1982.

50. **Bayer, E., Frank, H., and Nicholson, G.,** Carbamates were also used for enantiomer separation on Chirasil-Val, personal communication, 1982.

51. **Benecke, I. and König, W. A.,** Isocyanate als universelle reagentien bei der derivatbildung für die gaschromatographische enantiomerentrennung, *Angew. Chem.,* 94, 709, 1982; *Angew. Chem. Int. Ed. Engl.,* 21, 709, 1982; *Angew. Chem. Suppl.,* 1605, 1982.

52. **König, W. A., Benecke, I., Schulze, J., Sievers, S., Schmidt, E., and Lucht, N.,** Isocyanates as reagents for enantiomer separation: application to amino acids, N-methylamino acids and 3-hydroxy acids, *J. Chromatogr.,* 279, 555, 1983.

Chapter 2

COMPARISON OF D,L-AMINO ACID DERIVATIVES FOR RESOLUTION ON CHIRASIL-VAL® CAPILLARY COLUMNS AND AGE DATING OF SEDIMENTS

Iwao Abe and Shigefumi Kuramoto

TABLE OF CONTENTS

I. INTRODUCTION

Enantiomers of amino acids can be resolved directly by converting them to volatile *N*-perfluoroacyl alkyl ester derivatives with achiral reagents, followed by gas chromatography (GC) using a capillary column coated with Chirasil-Val®, which is an optically active stationary phase with high thermal stability.[1,2]

GC resolution of enantiomers has been carried out conventionally by derivatization to diastereoisomers with chiral reagents, followed by separation on achiral phases. This indirect method, in some instances, has advantages over the direct separation of enantiomers, but disadvantages exist.

The significant advantages and disadvantages of direct and indirect resolution of amino acids are as follows:

1. As commercial chiral reagents used in the diastereoisomeric method may not be highly pure, correction of peak areas is required. For the chiral phase method, peak area corrections are not required, with the exception of especially precise analyses in which the racemization of the solutes is a consideration.
2. A fractionation reaction, producing preferentially one of a pair of diastereoisomers, sometimes takes place between D,L-amino acids and a chiral reagent in the diastereoisomeric method. There is not the slightest fear to this problem in the chiral phase method using achiral reagents.
3. D-(+)-2-Butanol and *N*-trifluoroacetyl (TFA)-L-prolyl chloride are frequently used in the diastereoisomeric method.[3-8,9-16] As they are commercially available but extremely expensive, we have been obliged to use them sparingly, which limits experimental data.
4. Some diastereoisomers derived from *N*-TFA-L-prolyl chloride can be resolved on packed columns, but complete separation of mixtures of the protein amino acids is not possible.
5. Chiral phases give relatively large relative retention of all amino acids except proline and aspartic acid. Prior to the development of Chirasil-Val®, the thermal instability of earlier chiral phases necessitated lengthy chromatographic analyses. The use of Chirasil-Val® substantially reduces the problem of thermal instability of the chiral phase.

Direct resolution of amino acid enantiomers by gas-liquid chromatography (GLC) with chiral phase columns is both convenient and economical. This technique is going to take the place of the classical diastereoisomeric method.

The chiral phase method was first demonstrated by Gil-Av and co-workers.[17] In their pioneering work, several amino acids were resolved on glass capillary columns coated with *N*-TFA-L-isoleucine lauryl ester and *N*-TFA-L-phenylalanine cyclohexyl ester. Following their great work, attempts were made by many investigators to improve enantioselectivity and thermal stability.

Gil-Av prepared the first dipeptide phase, the well-known *N*-TFA-L-valyl-L-valine cyclohexyl ester in 1967, which greatly increased the resolution factor of all amino acids.[18] Immediately after, *N*-TFA-L-phenylalanyl-L-leucine cyclohexyl ester was synthesized by König et al., improving the maximum operating temperature to 140°C.[19]

The resolution mechanism of D,L-amino acids has been proposed as the formation of transient hydrogen-bonded association complexes between solutes and the chiral solvent. The resolution results from differences in the stability of the complexes between D- and L-amino acids and the chiral phase.[20-23] It has been noted that the best resolution could be obtained with amino acids esterified with a structurally bulky alcohol such as *t*-butanol and then perfluoroacylated.[20,24] Although amino acids esterified with *t*-butanol have given effective results, it is inconvenient to prepare this type of ester, and we were obliged to select isopropanol as a suitable esterifying reagent. In 1971, Feibush et al. reported a diamide phase of *N*-lauroyl-L-valine-*t*-butylamide with increased enantioselectivity and stability. This phase was later applied to the synthesis of the basic structure of Chirasil-Val®.[25]

Following this series of work, many phases were reported which used various high molecular weight amino acids of less volatility to improve thermal stability and resolution efficiency. These included *N*-TFA-L-valyl-L-leucine cyclohexyl ester and *N*-TFA-L-α-aminobutyryl-L-α-aminobutyric acid cyclohexyl ester by Parr et al.,[26,27] *N*-caproyl-L-valine *n*-hexylamide by Grohman et al.,[28] *N*-TFA L methionyl-L-methionine cyclohexyl ester by Andrawes et al.,[29] *N*-TFA-L-aspartic acid bis(cyclohexyl) ester by König et al.,[30] and *N*-caproyl-L-valyl-L-valine cyclohexyl ester by Abe et al.[31] A unique ureide-type phase of carbonyl-bis(L-valine isopropyl ester) was reported by Feibush et al. in 1967.[32] This phase very interestingly shows liquid crystal properties and gives extremely high resolution of amino acids to a certain extent at temperatures higher than the melting point.

In a novel approach, Charles et al.[33,34] modified the previously reported *N*-lauroyl-L-valine-*t*-butylamide by the introduction of extremely large *N*-acyl and/or amide moieties. *N,n*-Docosanoyl-L-valine-*t*-butylamide, *n*-lauroyl-L-valine-2-methyl-2-heptadecylamide, and *N-n*-docosanoyl-L-valine-2-methyl-2-heptadecylamide are stable for use up to 190, 180, and 200°C, respectively. Their enantioselectivities are sufficient for near complete resolution of all amino acids. These diamide phases are useful for practical analysis although basic amino acids such as lysine and tryptophan exhibit somewhat longer retention times.

After these pioneering research efforts, a new polysiloxane chiral phase was suddenly reported which fulfilled many research requirements. Frank and co-workers[35] first introduced a phase of this type, which possesses exceedingly low volatility and high thermal stability, by coupling L-valine-*t*-butylamide to a copolymer of dimethylpolysiloxane and carboxyalkylmethylsiloxane units. This development has resulted in the commercial availability of Chirasil-Val®. Chirasil-Val® capillary columns may be operated up to 220°C without appreciable bleeding which is especially important as the amino acid derivatives represent a wide range of molecular weights, vapor pressures, and polarities. Chirasil-Val® capillary columns provide excellent resolution, allow use of convenient temperature program, and withstand long periods of constant usage.[36] Chirasil-Val® capillary columns are also applicable to GC/mass spectroscopy (GC/MS) analysis, as they compare favorably with the normal achiral phases we have used in the past.[37]

In conventional cases, amino acids are converted to their *N(O,S)*-TFA or *N(O,S)*-PFP isopropyl ester derivatives to obtain suitable separation of the enantiomers. As described previously, the general concept has been known that the greater the bulkiness of alcohols used for esterification, the larger the relative retention. Therefore, isopropanol is most frequently used due to its reactivity and volatility. With respect to the acylating reagents, the more volatile pentafluoropropionyl (PFP) derivatives are reported to exhibit reduced retention times in comparison to the TFA or heptafluorobutyryl (HFB) derivatives.

In this paper, we report the GC resolution behavior of *N(O,S)*-perfluoracyl D,L-amino acid alkyl esters on Chirasil-Val®; the effect of the alcohols used for esterification as steric hindrance of bulky alkyl groups increases; and the influence of perfluoroacyl substituents on retention times, D and L peak-to-peak resolution, and elution order. Also, Heliflex™ Chirasil-Val®, a fused silica capillary column coated with Chirasil-Val®, which is the modified type of original Chirasil-Val®, is now commercially available.

This type of capillary further enhances thermal stability, decreases surface activity, and enhances ease of handling. The superiority of Heliflex™ Chirasil-Val® is demonstrated by means of our systematic investigation of various amino acid enantiomeric derivatives.

II. EXPERIMENTAL

A. Derivatization

1. D,L-*Amino Acid Stock Solution*

The standard D,L-amino acid stock solution containing the desired concentration of each amino acid must be prepared from reagent synthetic D- and L-amino acids; 6 μmol each of 19 protein D,L-amino acids are weighed and dissolved in 5 mℓ of 0.2 N HCl using a volumetric flask. Oxygen must be removed from the 0.2 N HCl with a gentle stream of nitrogen, both before and after the amino acids are dissolved. The flask is then stored in a refrigerator.

2. Preparation of HCl/Alcohol

a. Acetyl Chloride/Alcohol (20:80 v/v%)

For esterifications with methanol, ethanol, isopropanol, *n*-propanol, and neopentanol, acetyl chloride (up to 20 vol%) was added to each alcohol. Then 20 mℓ of chromatographically pure alcohol and a Teflon®-coated magnetic stirring bar are placed in a 50-mℓ Erlenmeyer® flask with a Teflon®-lined screw cap. After cooling the flask in an ice-water bath, 5 mℓ of acetyl chloride are added dropwise with continuous stirring.

b. HCl/Alcohol

When using acetyl chloride with the relatively high molecular weight alcohols such as isobutanol, *n*-butanol, and 3-pentanol, the HCl concentration tends to decrease with time from the initial concentration. Only in the case of neopentanol should the acetyl chloride/alcohol method be preferable, considering the neopentanol melting point of 53°C. Gaseous HCl from a compressed gas cylinder, 99.99% pure, is first passes through a concentrated H_2SO_4 trap to remove trace amounts of water, then the HCl is bubbled into 50 mℓ of alcohol in an Erlenmeyer® flask with a Teflon®-lined screw cap. The alcohol is continuously stirred with a Teflon®-coated magnetic stirring bar. The desired normality is obtained by withdrawing a 1.00-mℓ aliquot of the HCl/alcohol solution, mixing with 20 mℓ of water, and titrating with standardized NaOH solution. The final HCl concentrations are 3.5 N for isobutanol, 3.0 N for *n*-butanol, and 4.5 N for 3-pentanol, respectively.

3. Amino Acid Derivatives

Derivatives of the amino acid solutions are prepared by esterification and subsequent *N,O,S*-acylation as follows.

First, using a glass micropipet, 30 μℓ of the D,L-amino acid stock solution is placed in a microreaction vial, 1 mℓ volume with a Teflon®-lined septum. After placing this vial in a 100°C heat bath, the H_2O is removed slowly under a stream of dry nitrogen. A residual trace amount of H_2O can be removed azeotropically by the addition of 20 μℓ of methylene chloride after the vial has cooled to 50°C. Next, 100 μℓ of HCl/alcohol are added by micropipet and the vial is placed in an ultrasonic bath, 125W, for 30 sec to dissolve or suspend solid amino acids in the alcohol for subsequent heating at 100°C. For the methyl and ethyl esters, the heating period is 20 min; for isopropyl and *n*-propyl esters, 40 min; for isobutyl and *n*-butyl esters, 60 min; for neopentyl and 3-pentyl esters, 90 min. Of the various alcohols used for esterification, methanol is the most reactive, with 3-pentanol, from consideration of the molecular structure, least reactive. The excess reagents are then removed by evaporation under a gentle stream of dry nitrogen at 100°C, and the vial is cooled to room temperature. Next, 100 μℓ of methylene chloride is added, the sample is mixed in the ultrasonic bath for 30 sec, then the sample is cooled in an ice-water bath. The acylating reagent, trifluoroacetic anhydride (TFAA), pentafluoropropionic anhydride (PFPA), or heptafluorobutyric anhydride (HFBA) is then added by quickly pipetting 30 μℓ of the reagent into the vial. The vial is then tightly capped with a Teflon®-lined cap, and heated at 100°C for 10 min. After allowing the vial to cool, the excess reagents are evaporated under a gentle stream of dry nitrogen. Care must be exercised during the removal of these excess reagents in order to avoid loss by volatilization of derivatives, especially when using low molecular weight alcohols such as methanol, ethanol, and isopropanol in the esterification reaction. Finally, the derivatized amino acid mixture is dissolved in 50 $\mu\ell$ of ethyl acetate and 1.5- to 4-$\mu\ell$ aliquots are injected into the gas chromatograph.

III. RESULTS AND DISCUSSION

A. Effect of Alkyl and Perfluoroacyl Substituents

In order to investigate solute-solvent interactions in more detail, the behavior of *N(O)*-perfluoroacyl D,L-amino acid alkyl esters on Chirasil-Val® glass capillary columns was investigated. The effects of various alcohols used for esterification were studied, using increasingly more bulky alkyl groups, as similarly done in studies on the influence of substituents. The effects of the derivatization substituent groups were studied in terms of retention time, D and L peak-to-peak resolution, and elution order. In this study, amino acid mixtures were derivatized to the *N(O)*-perfluoroacyl (TFA, PFP, HFB) methyl, isopropyl, and 3-pentyl esters. A total of nine different derivatives were obtained.

Tables 1 to 3 show the retention times and D and L peak-to-peak resolution numbers (RN) of 17 *N(O)*- amino acid alkyl esters on Chirasil-Val® glass capillary columns.

The resolution numbers were calculated from the following equation:

$$RN = \frac{RT_L - RT_D}{W_{D/2} + W_{L/2}}$$

where RT_L is the retention time of the L-enantiomer, RT_D is the retention time of the D-enantiomer, $W_{D/2}$ is width at half-height of the D-enantiomer, and $W_{L/2}$ is width at half-height of L-enantiomer.

1. Effect of Perfluoroacyl Substituents on Resolution Numbers

Resolution numbers decrease on replacing TFA by PFP or HFB. This is a consequence of the decrease in solute-solvent hydrogen bond strengths with increasing size of the per-fluoroacyl group. This is very noticeable for the hydroxy amino acids threonine and serine.

2. Effect of Perfluoroacyl Substituents on Retention Times and Elution Order

Retention times generally decrease on replacing TFA by PFP or HFB. This trend is clearly seen for basic amino acids such as ornithine, lysine, and tryptophan. However, on replacing TFA by HFB, the retention times are slightly longer for alanine, valine, threonine, *allo*-isoleucine, glycine, leucine, proline, and serine among the more volatile amino acids, while they are shortened for aspartic acid, methionine, glutamic acid, and phenylalanine.

3. Effect of Alkyl Groups on Retention Times and Elution Order

The retention times of the amino acids increase with the size of the alkyl groups. The elution of aspartic acid and glutamic acid is strikingly delayed with increasing bulk of the alkyl substituents as these two amino acids are esterified at two points and the molecular weight consequently increases considerably.

Figure 1 shows a chromatogram of the *N(O)*-TFA methyl ester derivatives of a D,L-amino acid mixture. Although the early peaks elute quite closely, almost complete resolution is obtained for all these amino acids within 28 min. Peak overlap is observed only for L-threonine and D-*allo*-isoleucine, glycine, and D-isoleucine, and L-methionine and L-glutamic acid, respectively. Although it has been noted on occasion that the TFA-methyl ester de-

Table 1
RETENTION TIMES (RT) AND RESOLUTION NUMBERS (RN)
OF *N(O)*-PERFLUOROACYL AMINO ACID METHYL ESTERS

		TFA		PFP		HFB	
		RT(min)	RN	RT(min)	RN	RT(min)	RN
Ala	D	2.58	1.67	2.43	1.17	2.72	0.75
	L	2.78		2.57		2.90	
Val	D	3.21	1.75	2.92	1.15	3.34	0.85
	L	3.42		3.07		3.51	
Thr	D	3.90	NM	3.56	NM	4.38	NM
	L	4.28		3.77		4.50	
Gly		4.71	—	3.77	—	4.75	—
*a*Ile	D	4.28	NM	3.77	NM	4.50	NM
	L	4.71		4.09		4.75	
Ile	D	4.60	2.44	4.09	NM	4.75	NM
	L	4.99		4.34		5.01	
Leu	D	6.22	5.19	5.60	4.56	6.46	4.22
	L	7.05		6.33		7.22	
Pro	D	5.38	0.56	4.97	0.50	5.71	0
	L	5.50		5.09		5.71	
Ser	D	6.93	3.07	6.83	1.83	8.39	1.72
	L	7.36		7.16		8.70	
Met	D	13.11	NM	12.30	NM	13.15	NM
	L	13.79		12.96		13.71	
Asp	D	8.79	1.31	8.25	1.13	9.12	1.06
	L	9.00		8.43		9.29	
Phe	D	14.39	3.93	13.74	2.94	14.55	2.61
	L	14.94		14.21		15.02	
Glu	D	13.25	NM	12.44	NM	13.25	NM
	L	13.79		12.96		13.71	
Tyr	D	20.03	3.21	19.41	2.79	20.96	2.38
	L	20.48		19.80		21.34	
Orn	D	23.92	2.94	22.04	2.35	ND	—
	L	24.39		22.51			
Lys	D	25.39	1.85	23.56	1.75	ND	—
	L	25.76		23.91			
Trp	D	27.50	1.60	ND	—	ND	—
	L	27.82					

Note: NM: not measurable; ND: not detectable.

From Abe, I., Izumi, K., Kuramoto, S., and Musha, S., *HRC&CC,* 4, 549, 1981. With permission.

Table 2
RETENTION TIMES (RT) AND RESOLUTION NUMBERS (RN)
OF *N(O)* PERFLUOROACYL AMINO ACID ISOPROPYLESTERS

		TFA		PFP		HFB	
		RT(min)	RN	RT(min)	RN	RT(min)	RN
Ala	D	3.19	3.42	3.08	3.00	3.43	1.89
	L	3.60		3.38		3.77	
Val	D	4.24	2.81	3.96	2.67	4.47	1.83
	L	4.69		4.28		4.80	
Thr	D	4.80	NM	4.63	2.33	5.74	NM
	L	5.36		4.91		5.92	

Table 2 (continued)
RETENTION TIMES (RT) AND RESOLUTION NUMBERS (RN) OF *N(O)* PERFLUOROACYL AMINO ACID ISOPROPYLESTERS

			TFA		PFP		HFB	
			RT(min)	RN	RT(min)	RN	RT(min)	RN
Gly			5.36	—	5.32	—	6.06	—
*a*Ile	D		5.54	4.86	5.14	3.40	5.74	NM
	L		6.22		5.65		6.20	
Ile	D		5.95	4.26	5.51	3.36	6.20	NM
	L		6.55		5.98		6.57	
Leu	D		7.34	9.58	6.92	NM	7.57	6.31
	L		8.49		7.93		8.58	
Pro	D		7.05	0.50	6.78	NM	7.43	0
	L		7.14		6.92		7.43	
Ser	D		7.76	4.08	7.93	NM	9.27	1.53
	L		8.29		8.26		9.53	
Met	D		14.20	5.79	13.67	4.77	14.22	3.76
	L		15.01		14.34		14.86	
Asp	D		12.20	2.15	11.94	1.17	12.45	0.69
	L		12.48		12.08		12.56	
Phe	D		15.75	4.71	15.28	3.93	15.82	3.18
	L		16.41		15.83		16.36	
Glu	D		16.18	3.32	15.60	2.94	16.03	2.83
	L		16.81		16.13		16.54	
Tyr	D		20.63	3.29	20.13	2.88	21.42	2.64
	L		21.19		20.59		21.87	
Orn	D		24.15	3.00	22.58	2.88	23.40	2.58
	L		24.69		23.07		23.84	
Lys	D		25.79	2.44	24.23	2.11	25.01	1.83
			26.23		24.61		25.34	
Trp	D		27.79	1.79	26.61	1.50	ND	—
	L		28.22		26.85			

Note: NM: not measurable; ND: not detectable.

From Abe, I., Izumi, K., Kuramoto, S., and Musha, S., *HRC&CC,* 4, 549, 1981. With permission.

Table 3
RETENTION TIMES (RT) AND RESOLUTION NUMBERS (RN) OF *N(O)*-PERFLUOROACYL AMINO ACID 3-PENTYL ESTERS

			TFA		PFP		HFB	
			RT(min)	RN	RT(min)	RN	RT(min)	RN
Ala	D		6.76	4.95	6.38	4.50	7.17	4.06
	L		7.70		7.10		7.90	
Val	D		8.51	NM	8.02	2.00	8.84	1.69
	L		9.05		8.38		9.17	
Thr	D		9.05	NM	8.60	1.75	10.11	1.31
	L		9.56		8.88		10.32	
Gly			9.97	—	9.75	—	10.64	—
*a*Ile	D		9.97	NM	9.39	2.63	10.17	NM
	L		10.64		9.89		10.64	
Ile	D		10.64	NM	9.99	2.47	10.73	2.25
	L		11.11		10.36		11.09	

Table 3 (continued)
RETENTION TIMES (RT) AND RESOLUTION NUMBERS (RN) OF *N(O)*-
PERFLUOROACYL AMINO ACID 3-PENTYL ESTERS

		TFA		PFP		HFB	
		RT(min)	RN	RT(min)	RN	RT(min)	RN
Leu	D	11.90	7.19	11.20	NM	11.92	6.56
	L	13.05		12.25		12.97	
Pro	D	12.10	0.33	11.58	0.29	12.21	0
	L	12.16		11.65		12.21	
Ser	D	12.35	3.00	12.25	NM	13.44	1.43
	L	12.80		12.44		13.64	
Met	D	18.64	4.17	17.82	3.61	18.32	2.63
	L	19.39		18.47		18.94	
Asp	D	20.32	NM	19.76	0	20.05	NM
	L	20.51		19.76		20.05	
Phe	D	20.32	NM	19.56	2.67	20.05	NM
	L	20.85		20.04		20.49	
Glu	D	24.24	2.89	23.33	NM	23.57	2.50
	L	24.76		23.78		24.02	
Tyr	D	24.53	2.82	23.78	NM	24.81	2.28
	L	25.01		24.14		25.22	
Orn	D	27.73	2.58	25.79	2.45	26.42	2.25
	L	28.35		26.28		26.87	
Lys	D	29.95	2.20	27.51	2.08	28.18	1.88
	L	30.61		27.93		28.63	
Trp	D	33.05	2.11	30.43	1.95	30.89	1.40
	L	33.81		30.86		31.32	

Note: NM: not measurable.

From Abe, I., Izumi, K., Kuramoto, S., and Musha, S., *HRC&CC*, 4, 549, 1981. With permission.

rivatives of amino acids are too volatile for practical use, because of the severe losses of low molecular weight amino acids such as glycine, alanine, valine, etc., during removal by vaporization of the reaction solvent and excess reagents, it appears that these derivatives are indeed useful, given rigid control of the temperature and the time of vaporization in the course of the derivatization process.

Figure 2 shows a chromatogram of the *N(O)*-TFA isopropyl esters of an amino acid mixture. This type of derivative has been very well known from the beginning of the direct resolution method. It is well known that isopropyl ester derivatives are well resolved. Enantiomer separation is complete for all amino acids except proline. It is overlapped by threonine and glycine, with slight overlap of L-valine and D-threonine. The retention time of the final solute, L-tryptophan, is within 29 min and is not significantly greater in comparison with the *N(O)*-TFA methyl ester derivatives. TFA isopropyl ester derivatives have been widely applied to the GC resolution of amino acids, and is a practical analytical method.

Figure 3 shows a chromatogram of the *N(O)*-TFA 3-pentyl ester derivatives of a standard amino acid mixture. In the earliest stages of this area of research, this type of derivative had been said to yield very high resolution due to an increase in the effects of steric enantioselective interaction. Almost all amino acids show their largest RN with this derivative which can be easily appreciated in Figure 3 and Table 3.

Unfortunately, peak overlapping occurred more than was observed in Figures 1 or 2. L-Valine and D-threonine, glycine and D-*allo*-isoleucine, D- and L-proline, and D-phenylalanine and D-aspartic acid essentially coelute.

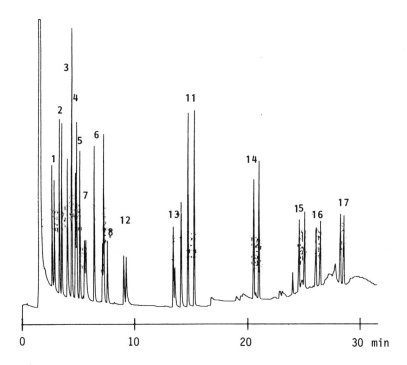

FIGURE 1. Chromatogram of TFA-methyl derivatives of DL-amino acid mixture. Column conditions: 25 m × 0.3 mm Chirasil-Val® glass capillary column; temperature, 90°C; 4-min hold, then programmed to 180°C at 4°/min; injector temperature, 250°C; carrier gas, helium; split ratio, 1:35. (1) Ala; (2) Val; (3) Thr; (4) aIle; (5) Ile; (6) Leu; (7) Pro; (8) Ser; (9) L-Hyp; (10) Met; (11) Phe; (12) Asp; (13) Glu; (14) Tyr; (15) Orn; (16) Lys; (17) Trp. For each pair of peaks the D-enantiomer emerges faster.

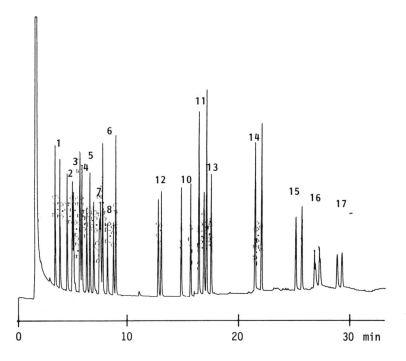

FIGURE 2. Chromatogram of TFA-isopropyl derivatives of DL-amino acid mixture. Conditions and peak identities are shown in Figure 1.

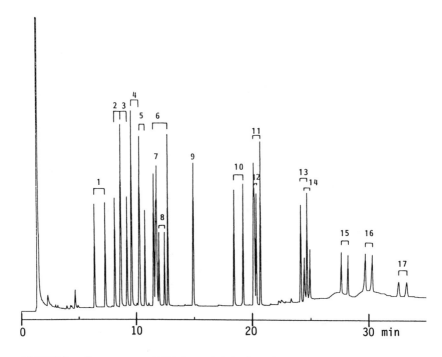

FIGURE 3. Chromatogram of TFA-3-pentyl derivatives of DL-amino acid mixture. Conditions and peak identities are shown in Figure 1. (From Abe, I., Izumi, K., Kuramoto, S., and Musha, S., *HRC&CC,* 4, 459, 1981. With permission.

It should be said that 3-pentyl ester derivatives may be useful for the relatively volatile amino acids, with the exceptions of valine and isoleucine because of their reduced reactivity with 3-pentanol. The large α-alkyl substituents of these amino acids presumably interfere in the esterification reaction.

IV. HELIFLEX™ CHIRASIL-VAL® CAPILLARY COLUMNS

Fused silica has proven to be an ideal material for capillary column construction, being inherently more inert than any other glasses. Also, the flexibility and mechanical durability of fused silica capillary columns make them easy to handle and install in column ovens. The troublesome problem of straightening column ends associated with conventional glass capillary columns is eliminated.

In 1982, a modified type of Chirasil-Val® was reported by Wulff incorporating 15% phenyl groups substituted for methyl groups in the dimethylsiloxane units to improve thermal stability and separation efficiency.[38] This new phase of slightly modified Chirasil-Val®, coated on fused silica capillary columns, is marketed as Heliflex™ Chirasil-Val® (Applied Science Division, Milton Roy Co., State College, Pa.) U.S.

In view of the potential of this new type of column, we made a decision to consider this column with regard to our investigations of the GC behavior of various enantiomeric derivatives of amino acid mixtures.

We prepared nine different derivatives of standard amino acid mixtures; TFA methyl esters, TFA ethyl esters, TFA isopropyl esters, PFP isopropyl esters, PFP *n*-propyl esters, TFA isobutyl esters, TFA *n*-butyl esters, TFA neopentyl esters, and TFA 3-pentyl esters, to investigate and elucidate the behavior of D and L resolution and peak-to-peak separation as this had not yet been reported for Heliflex™ Chirasil-Val® in detail.

A Heliflex™ Chirasil-Val® column, 25 m in length, 0.31 mm i.d., with 0.23 μm film

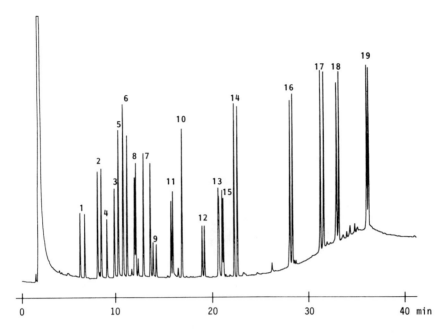

FIGURE 4. Chromatogram of TFA-methyl derivatives of DL-amino acid mixture. Column conditions: 25 m × 0.31 mm fused silica Chirasil-Val® capillary column; temperature, 80°C; 4-min hold, then programmed to 200°C at 4°/min; injector temperature, 200°C; carrier gas, helium; split ratio, 1:30. (1) Ala; (2) Val; (3) Thr; (4) Gly; (5) *a*Ile; (6) Ile; (7) Leu; (8) Pro; (9) Ser; (10) L-Hyp; (11) Asp; (12) Cys; (13) Met; (14) Phe; (15) Glu; (16) Tyr; (17) Orn; (18) Lys; (19) Trp. For each pair of peaks the D-enantiomer emerges faster.

thickness, was obtained from Applied Science Division, mounted in a GC, and conditioned at 210°C for 24 hr after gradually raising the oven temperature at 4°/min.

N(O,S)-**TFA methyl esters** — These derivatives are the most volatile of all derivatives prepared. Figure 4 shows a chromatogram of the *N(O,S)*-TFA methyl esters of a D,L-amino acid mixture. The temperature program was started at 80°C, with a 4-min hold, and then raised at 4°/min to 200°C for these highly volatile, low molecular weight methyl ester derivatives. It was observed that almost all amino acids could be resolved as their enantiomers, the D and L enantiomers separated from each other quite well. Only D-methionine and D-glutamic acid overlapped. Proline shows the best resolution of any other derivatives described later, also aspartic acid elutes very fast and is resolved completely. It has been reported that methyl ester derivatives of amino acids are too volatile for practical use, have a tendency to form peptide bonds at high temperatures during the derivatizing process, and also have low relative molar responses (RMR) with flame ionization detection (FID). On the other hand, the high volatility enables these derivatives to elute at low temperatures; thus high resolution can be obtained in a fairly short time.

N(O,S)-**TFA ethyl esters** — The ethyl ester derivatives are seldom used, probably because of reported problems of loss of low molecular weight amino acid derivatives such as alanine, valine, glycine, isoleucine, and leucine during evaporation of excess reagents and solvent, similar to the case of the methyl ester derivatives. Like methyl esters, the high volatility of the ethyl esters allow elution at relatively low column temperatures with high resolution. A chromatogram of the *N(O,S)*-TFA ethyl esters of an amino acid mixture is shown in Figure 5. The clear separation of four enantiomeric peaks, D, L-glutamic acid and D,L-phenylalanine, might catch your attention first, and other amino acids are resolved almost completely. Overlapping is observed for L-*allo*-isoleucine and D-leucine, D-serine and L-leucine, also slightly for D- and L-proline. It would be said that this derivative is best on Heliflex™

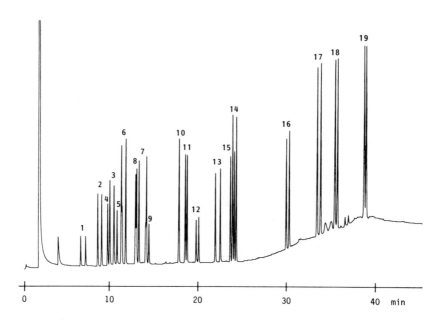

FIGURE 5.   Chromatogram of TFA-ethyl derivatives of DL-amino acid mixture. Temperature, 85°C; 4-min hold, then programmed to 200°C at 3.5°/min. Other conditions and peak identities are shown in Figure 4.

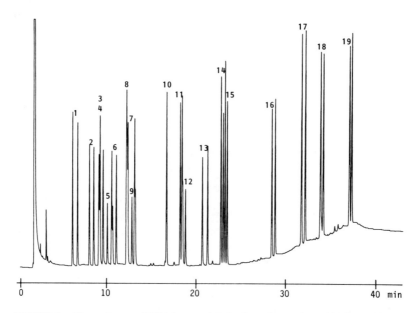

FIGURE 6.   Chromatogram of TFA-isopropyl derivatives of DL-amino acid mixture. Temperature, 90°C; 4-min hold, then programmed to 200°C at 3.5°/min. Other conditions and peak identities are shown in Figure 4.

Chirasil-Val® if loss of derivatives are avoided during vaporization of reagents and solvent prior to GC analysis.

N(O,S)-TFA isopropyl esters — Figure 6 is a chromatogram of the *N(O,S)*-TFA isopropyl esters of an amino acid mixture. It has been reported that these derivatives have high RN on all chiral phases. As can be seen from Figure 6, almost all amino acids can be resolved with some exceptions. Enantiomeric pairs of D,L-phenylalanine and D,L-glutamic acid are

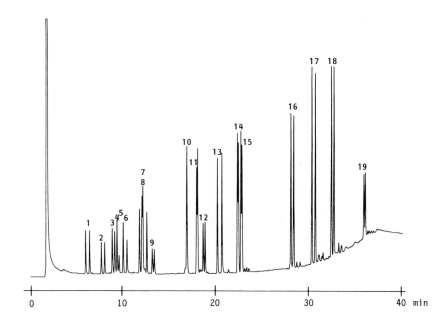

FIGURE 7.   Chromatogram of PFP-isopropyl derivatives of DL-amino acid mixture. Temperature, 90°C, 4-min hold, then programmed to 200°C at 3.5°/min. Other conditions and peak identities are shown in Figure 4.

all separated completely. Overlapping is observed for D-valine and glycine, D-isoleucine and L-*allo*-isoleucine, D,L-proline and D-leucine, L-leucine and L-serine, and L-aspartic acid and D-cysteine, respectively.

N(O,S)-PFP isopropyl esters — Figure 7 is a chromatogram of the *N(O,S)*-PFP isopropyl amino acid esters. The PFP derivatives are all more volatile than the TFA derivatives. This is clearly observed by comparison of Figure 7 and Figure 6, noting especially the basic amino acids, such as ornithine, lysine, and tryptophan. As shown, Figure 7 resembles somewhat Figure 6 in appearance, and almost all amino acids are resolved completely. However, D,L-phenylalanine and D,L-glutamic acid, D- and L-proline and D- and L-aspartic acid overlap.

N(O,S)-PFP n-propyl esters — Figure 8 shows the *N(O,S)*-PFP *n*-propyl esters of a standard amino acid enantiomeric mixture. This type of derivative has been reported by Frank et al. using Chirasil-Val® as the stationary phase, as practical for enantiomeric analysis.[52] All amino acids elute without overlapping each other, but resolution was incomplete for D- and L-proline and for D- and L-aspartic acid.

N(O,S)-TFA isobutyl esters — Figure 9 shows the unique chromatographic behavior of the TFA isobutyl ester derivatives, which somewhat resembles Figure 8. Nearly all amino acids are found to elute at suitable intervals without serious overlapping. The resolution is efficient for the volatile amino acids; thus alanine, valine, threonine, glycine, and isoleucine were well resolved. Less volatile amino acids, cysteine, methionine, phenylalanine, glutamic acid, tyrosine, ornithine, lysine, and tryptophan were also well resolved, with proline, serine, and aspartic acid showing some overlap or incomplete resolution. Using this derivative, neighboring peaks are nicely arranged at nearly regular intervals.

N(O,S)-TFA n-butyl esters — In this case, less volatile amino acids, such as tyrosine, ornithine, and tryptophan, show complete resolution, as does alanine, valine, threonine, glycine, cysteine, and methionine, as shown in Figure 10. However, D,L-proline has overlapped with L-leucine and D-serine. Also D,L-aspartic acid has emerged just between D- and L-phenylalanine.

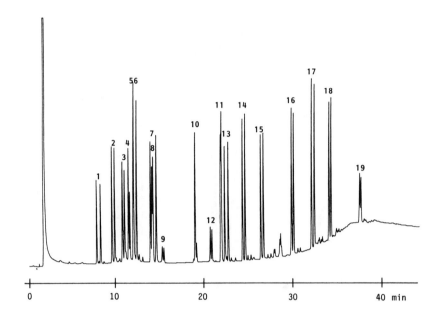

FIGURE 8.   Chromatogram of PFP-*n*-propyl derivatives of DL-amino acid mixture. Temperature, 90°C, 4-min hold, then programmed to 200°C at 3.5°/min. Other conditions and peak identities are shown in Figure 4.

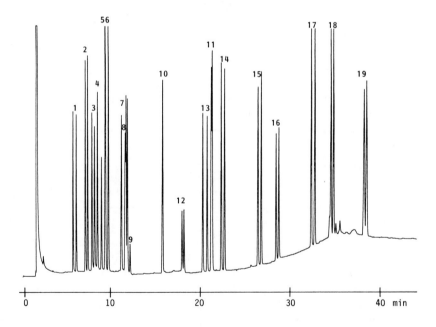

FIGURE 9.   Chromatogram of TFA-isobutyl derivatives of DL-amino acid mixture. Temperature, 110°C, 4-min hold, then programmed to 200°C at 3°/min. Other conditions and peak identities are shown in Figure 4.

N(O,S)-**TFA neopentyl esters** — As we found no report of this derivative, we were particularly interested in this study. Neopentanol is reported to be unstable in the presence

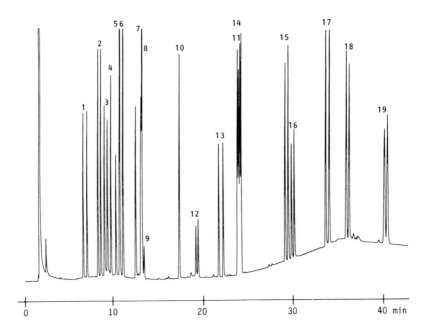

FIGURE 10. Chromatogram of TFA-*n*-butyl derivatives of DL-amino acid mixture. Temperature, 110°C; 4-min hold, then programmed to 200°C at 3°/min. Other conditions and peak identities are shown in Figure 4.

of an acid catalyst, having the property of forming stable *t*-amyl derivatives by a rearrangement reaction as follows:

$$CH_3-\underset{\underset{CH_3}{|}}{\overset{\overset{CH_3}{|}}{C}}-CH_2OH \xrightarrow{H^+} CH_3-\underset{\underset{CH_3}{|}}{\overset{\overset{CH_3}{|}}{C}}-CH_2^+ \rightarrow CH_3-\underset{\underset{CH_3}{|}}{C^\pm}CH_2CH_3$$

As part of this experiment, considerable peak identification was required. As shown in Figure 11, alanine, valine, methionine, glutamic acid, tyrosine, ornithine, and lysine are well resolved. However, this type of derivative is not considered necessarily suitable for the resolution of amino acids.

N(O,S)-**TFA 3-pentyl esters** — *N(O,S)*-TFA 3-pentyl derivatives have the interesting feature of separating four isoleucine enantiomers, as shown in Figure 12, which could not be observed with other derivatives. The low relative peaks of these enantiomers may result from reduced reactivity with 3-pentanol. The chromatogram is not so satisfactory but the resolution (RN) is large enough to resolve all amino acids except proline and aspartic acid. L-Leucine and D,L-proline, aspartic acid and D-phenylalanine, and L-tyrosine and D-ornithine overlap to varying degrees.

The 9 chromatograms described above, with distinguishing features in some respects, have been selected from our studies of 22 derivatives. None of these derivatization procedures resulted in complete resolution of all the amino acid enantiomers in the amino acid standard solution in a single chromatographic analysis. Complete separation might be possible as the result of careful investigations of the effects of various chromatographic conditions, such as programming rate, gas flow rate, and sample amount. However, if these amino acids are derivatized to two or three different types of derivatives, and chromatographed individually

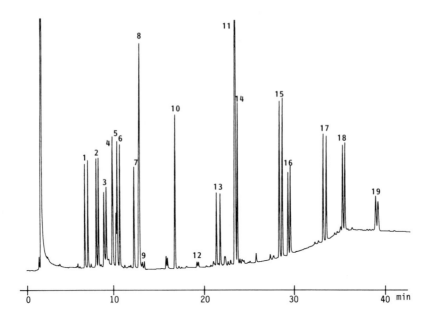

FIGURE 11.   Chromatogram of TFA-neopentyl derivatives of DL-amino acid mixture. Temperature, 110°C, 4-min hold, then programmed to 200°C and 3°/min. Other conditions and peak identities are shown in Figure 4.

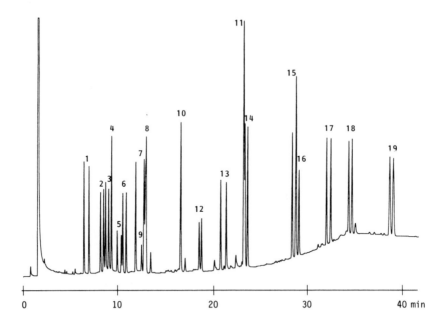

FIGURE 12.   Chromatogram of TFA-3-pentyl derivatives of DL-amino acid mixture. Temperature, 110°C; 4-min hold, then programmed to 200°C at 3°/min. Other conditions and peak identities are shown in Figure 4.

on Heliflex™ Chirasil-Val®, then all these amino acids are completely resolved. We can divide these 19 amino acids into two classes: one class is the relatively volatile amino acids of alanine, valine, threonine, glycine, *allo*-isoleucine, isoleucine, proline, leucine, and

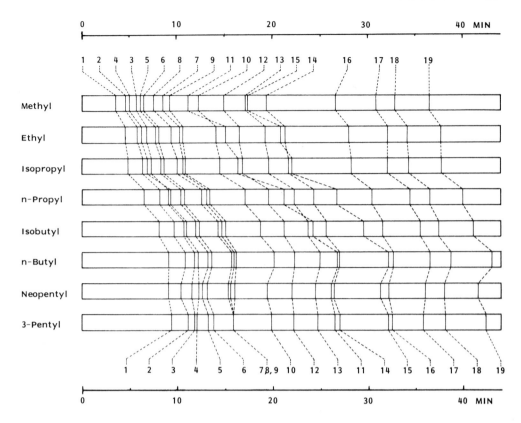

FIGURE 13.   Elution characteristics of various TFA amino acid esters on Heliflex™ Chirasil-Val®. Temperature, 100°C; 4-min hold and then programmed at 3°/min to 200°C. (1) L-Ala; (2) L-Val; (3) L-Thr; (4) Gly; (5) L-*a*Ile; (6) L-Ile; (7) L-Leu; (8) L-Pro; (9) L-Ser; (10) L-Hyp; (11) L-Asp; (12) L-Cys; (13) L-Met; (14) L-Phe; (15) L-Glu; (16) L-Tyr; (17) L-Orn; (18) L-Lys; (19) L-Trp. Other conditions are shown in Figure 4. (From Abe, I., Kuramoto, S., and Musha, S., *HRC&CC,* 6, 366, 1983. With permission.)

serine; the other consists of the less volatile amino acids, hydroxyproline, cysteine, aspartic acid, methionine, phenylalanine, glutamic acid, tyrosine, ornithine, lysine, and tryptophan, and discuss the derivatives which yield best resolution with regard to the chromatograms described in Figures 1 to 9. First, the best resolution of the volatile amino acids is produced by the *N(O,S)*-PFP isopropyl esters as shown in Figure 4. However, four components overlap (L-*allo*-isoleucine and D-isoleucine and D- and L-proline), while the others show excellent resolution. The less volatile amino acids are well resolved as the *N(O,S)*-TFA ethyl esters as shown in Figure 2. The possibility of losses during the vaporization processes should not necessarily influence researchers to use less volatile solutes. Efficient resolution can be obtained for aspartic acid, and also for the enantiomeric pairs of glutamic acid and phenylalanine, which are interspersed.

Finally, the elution order of all derivatives was investigated by classifying them into three groups: TFA, PFP, and HFB derivatives.

Figures 13 to 15 show a schematic representation of the retention times of 19 L-amino acid esters of TFA, PFP, and HFB derivatives on Heliflex™ Chirasil-Val®. In general, retention times increased with increasing size of alkyl ester substituent. This is clearly observed from the methyl to isobutyl esters. However, the *n*-butyl esters were inclined to be eluted slightly later than either the neopentyl or 3-pentyl esters, with the neopentyl esters being eluted earlier than the 3-pentyl esters.

Therefore, the order of emergence has been arranged as methyl, ethyl, isopropyl, *n*-propyl, isobutyl, neopentyl, 3-pentyl, and *n*-butyl esters.

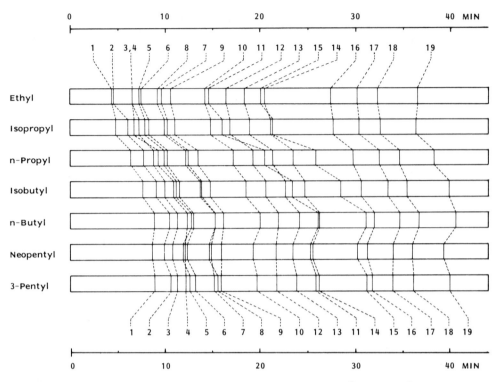

FIGURE 14.   Elution characteristics of various PFP amino acid esters on Heliflex™ Chirasil-Val®. Other conditions and peak identities are shown in Figure 13. (From Abe, I., Kuramoto, S., and Musha, S., *HRC&CC*, 6, 366, 1983. With permission.)

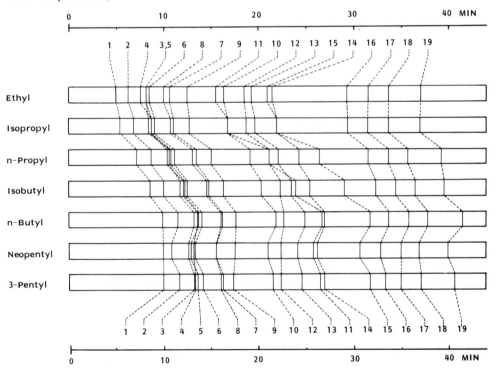

FIGURE 15.   Elution characteristics of various HFB amino acid esters on Heliflex™ Chirasil-Val®. Other conditions and peak identities are shown in Figure 13. (From Abe, I., Kuramoto, S., and Musha, S., *HRC&CC*, 6, 366, 1983. With permission.)

N(O,S)-TFA *n*-propyl esters might be suitable for separation of amino acid mixtures to a reasonable extent, but their resolution is incomplete.

With regard to the neutral and acidic amino acids, retention times increase from PFP to TFA to HFB acyl groups, and for the basic amino acids, PFP to HFB to TFA, derivatives.

As previously described, the retention times of aspartic acid and glutamic acid are considerably lengthened with increasing size of the alkyl substituent with which esterified. For example, the retention time of *N*-TFA-aspartic acid bis *n*-butyl ester is about 2.4-fold longer than for the methyl ester.

The major reason that all the amino acid enantiomers cannot be separated in a single run is that the lower molecular weight amino acids elute during a narrow interval. As seen in Figure 15, threonine, glycine, *allo*-isoleucine, and isoleucine are eluted within a very narrow range.

Table 4 lists the resolution factors for 19 protein amino acids. The conditions are the same as described in Figures 13 to 15.

V. APPLICATION TO AGE-DATING OF FOSSIL SHELLS IN SEDIMENTS

Amino acids, indispensable to living organisms, constitute various forms of peptides and proteins. As the protein amino acids are of the α-amino acid form, there are inevitably optical isomers of each. However, the wonder is that all protein amino acids are L-enantiomers only, a phenomenon the origin of which has not yet been thoroughly explained. Amino acids, especially in proteins, are said to be very stable if left as they are. In 1968, Hare and Abelson first found D-enantiomers in fossil shells in their research on the distribution and transformation of amino acids and proteins over geological time.[39]

The amino acid racemization dating method was first suggested by Hare and Mitterer in 1969, in accordance with their outstanding work on the possibility of applying the epimerization of L-isoleucine to D-*allo*-isoleucine to geochronology.[40] This technique has been applied to fossil shells, bones, trees, and deep-sea sediments.[41-48] We also found that the determination of aspartic acid racemization could date tree rings of Yakusugi (*Cryptomeria japonica*).[46] As mentioned earlier, the most important problem is to evaluate the temperature to which the sample has been exposed. This problem has been approached from several directions.[49,50]

The total amino acid composition of the sample and the extent of epimerization of isoleucine to *allo*-isoleucine can also be determined on an automatic amino acid analyzer.[42] However, the D:L ratios of amino acids could only be determined by GC. GC is suitable for resolving amino acids in view of the high resolution efficiency, especially of Chirasil-Val® capillary columns, speed, and high sensitivity and precision. Chirasil-Val® capillary columns resolve almost all protein amino acids in a short analysis time, and have been used extensively in practical analysis.[51,52]

Here, we discuss a recent application of amino acid racemization dating to sediments at Osaka South Port along Osaka Bay.

Osaka Plain, Japan, a region where crocodile fossils were discovered in 1964, has attracted the geochemists' attention regarding cultural anthropology in the Pleistocene age. Many fossil bones and teeth of *Paleoloxodon naumanni* and *Parastegodon akashensis* several hundred thousand years of age were discovered successively, proving that Japan was once contiguous to the Continent of Asia. It has been assumed that the formation of the present Osaka Plain was caused by upheaval at the bottom of the sea due to a few transgressions and regressions from the period of Upper Pleistocene to Holocene.[53] The presumption may be verified from the age-dating of sediments at various depths. However, there are two problems in deducing the ages of the sediments by directly estimating the extent of the racemization of amino acids. One is the troublesome initial determination of the amounts

Table 4
RESOLUTION FACTORS[a] FOR 19 PROTEIN AMINO ACIDS AS THEIR PERFLUOROACYL ALKYL ESTER DERIVATIVES ON HELIFLEX™ CHIRASIL-VAL®

No.	Amino acid	Me	Et			iPr			nPr			iBu			nBu			neoPe			3-Pe		
		TFA	TFA	PFP	HFB	TFA	PFP	HFB	TFA	PFP	HFB	TFA	PFP	HFB	TFA	PFP	HFB	TFA	PFP	HFB	TFA	PFP	HFB
1	Ala	1.060	1.080	1.062	1.057	1.089	1.065	1.064	1.081	1.054	1.066	1.052	1.045	1.049	1.054	1.040	1.045	1.040	—	—	1.080	1.067	1.058
2	Val	1.046	1.056	1.042	1.036	1.065	1.045	1.041	1.047	1.035	1.037	1.029	1.023	1.025	1.030	1.024	1.019	—	—	—	1.040	1.026	1.022
3	Thr	1.050	1.054	1.035	1.017	1.036	1.038	1.024	1.033	1.027	1.037	1.021	1.015	1.018	1.029	1.016	1.014	—	—	—	1.035	1.018	1.015
4	Gly	—	—	—	—	—	—	—	—	—	—	—	—	—	—	—	—	—	—	—	—	—	—
5	aIle	1.058	1.055	1.051	1.048	1.067	1.053	1.048	1.048	1.043	1.039	1.039	1.031	1.039	1.034	1.028	1.025	1.035	—	—	1.040	—	—
6	Ile	1.049	1.052	1.043	1.037	1.057	1.045	1.041	1.035	1.041	1.037	1.033	1.032	1.023	1.027	1.027	1.021	1.019	—	—	1.031	1.024	1.020
7	Leu	1.081	1.086	1.072	1.069	1.093	1.082	1.076	1.063	1.058	1.054	1.051	1.048	1.044	1.044	1.040	1.038	—	—	—	1.066	1.059	1.056
8	Pro	1.015	1.013	1.011	1.007	1.007	1.010	1.005	1.010	1.009	1.008	1.009	1.005	1.005	1.007	1.002	1.000	—	—	—	1.005	1.003	1.000
9	Ser	1.037	1.031	1.023	1.018	1.039	1.022	1.015	1.019	1.017	1.011	1.012	1.010	1.016	1.015	1.008	1.008	—	—	—	1.020	1.011	1.010
10	L-Hyp	—	—	—	—	—	—	—	—	—	—	—	—	—	—	—	—	—	—	—	—	—	—
11	Asp	1.017	1.015	1.010	1.008	1.014	1.008	1.004	1.005	1.006	1.003	1.004	1.002	1.005	1.002	1.000	1.000	—	—	—	1.004	1.002	1.000
12	Cys	1.020	1.020	1.015	1.011	1.023	1.015	1.010	1.011	1.011	1.008	1.008	1.006	1.012	1.008	1.006	1.005	—	—	—	1.012	1.008	1.006
13	Met	1.031	1.034	1.029	1.025	1.036	1.031	1.027	1.026	1.022	1.020	1.018	1.017	1.020	1.017	1.016	1.015	—	—	—	1.025	1.022	1.020
14	Phe	1.024	1.023	1.019	1.018	1.025	1.021	1.019	1.018	1.015	1.015	1.012	1.011	1.015	1.011	1.011	1.011	—	—	—	1.015	1.013	1.011
15	Glu	1.035	1.026	1.022	1.020	1.025	1.022	1.021	1.016	1.014	1.013	1.012	1.011	1.012	1.011	1.010	1.010	—	—	—	1.013	1.012	1.011
16	Tyr	1.014	1.015	1.012	1.011	1.016	1.013	1.011	1.011	1.010	1.009	1.008	1.008	1.010	1.008	1.007	1.006	—	—	—	1.010	1.010	1.008
17	Orn	1.015	1.015	1.014	1.012	1.015	1.014	1.013	1.012	1.011	1.011	1.011	1.009	1.011	1.009	1.008	1.008	—	—	—	1.012	1.010	1.010
18	Lys	1.010	1.011	1.009	1.008	1.011	1.010	1.008	1.009	1.008	1.007	1.006	1.006	1.008	1.007	1.006	1.005	—	—	—	1.010	1.012	1.007
19	Trp	1.006	1.006	1.005	1.005	1.008	1.007	1.004	1.004	1.004	1.007	1.004	1.004	1.008	1.005	1.005	1.006	—	—	—	1.009	1.006	1.005

[a] Resolution factor: $r = t_L/t_D$ where t_L and t_D are the retention times of L- and D-enantiomers.

From Abe, I., Kuramoto, S., and Musha, S., *HRC&CC*, 6, 366, 1983. With permission.

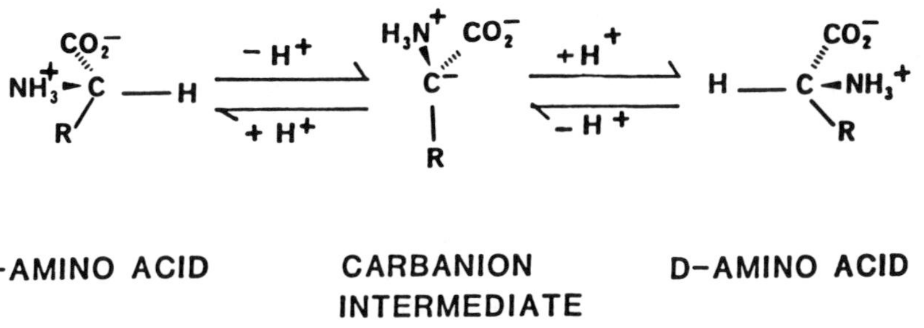

L–AMINO ACID **CARBANION INTERMEDIATE** **D–AMINO ACID**

FIGURE 16. Mechanism of amino acid racemization.

of D-amino acids in the sediments, which vary remarkably with the ages of the formed sediments. The other is the complicated fractionation process of the sediments with regard to the various forms of amino acids, i.e., proteins, peptides, and free amino acids, which have an effect on the rates of amino acid racemization. The fractionation to free and bound amino acids is a significant factor in the dating of sediments. These problems could be solved by determining the ages of the shells contained in the sediments. A single analysis requires about 0.2 g of shells to maintain precision and accuracy. A novel procedure for age-dating is described.

A. Amino Acid Racemization

A racemization mechanism of amino acids has been proposed by Smith and Evans as shown in Figure 16.[54] This mechanism is applicable to not only "free" amino acids but also "peptide" and "protein" amino acids.[55] The racemization of L-amino acids into D-amino acids is a reversible first-order reaction. The reaction can be written as follows:

$$\text{L-Amino Acid} \underset{k_D}{\overset{k_L}{\rightleftharpoons}} \text{D-Amino Acid} \tag{1}$$

where k_L and k_D are the racemization reaction rate constants of L- and D-amino acids, respectively, This reaction continues until equal amounts of both enantiomers are eventually produced. The rate of racemization of an L-amino acid can be written as:

$$\frac{-d[\text{L}]}{dt} = k_L[\text{L}] - k_D[\text{D}] \tag{2}$$

where [L] and [D] are the concentrations of L- and D-amino acids, respectively. Integration of Equation 2 yields:[48]

$$\ln\left[\frac{1 + (\text{D/L})}{1 - K'(\text{D/L})}\right]_t - \ln\left[\frac{1 + (\text{D/L})}{1 - K'(\text{D/L})}\right]_{t_0} = (1 + K')k_L t \tag{3}$$

where $K' = k_D/k_L$, and D:L is the ratio of amino acid enantiomers. The t=0 term is the correction due to racemization during the analytical procedure. As all these terms could be obtained from experimental data, the ages of the fossils could be obtained from Equation 3. The K' value is not fixed, but depends on the number of asymmetric carbon atoms. For amino acids with a single asymmetric center, e.g., alanine, valine, and aspartic acid, K' = 1. For amino acids with two or more asymmetric centers e.g., isoleucine, $K' \neq 1$.[56]

As racemization is a chemical reaction, the rate constant depends on the surrounding environment, such as temperature, pH, and ionic strength.[50,57-59] Temperature especially affects the rate of racemization. For example, at 0°C, racemization half-lives are about 100 times that at 25°C.[50] Therefore, knowledge of the thermal history of the fossils is required. The rate is also dependent on the structure of the amino acids. Amino acids are reported to be racemized in the following order; leucine $<$ isoleucine $<$ glutamic acid $=$ alanine $<$ aspartic acid.[44] Aspartic acid racemization is used to date fossils of less than about a few ten thousands of years old. In older fossils, alanine racemization or isoleucine epimerization has been used.[45] Following is a representation of fossil age dating.

Several fossils are collected from the same region; the absolute age of at least one fossil must be determined by a method such as radiocarbon dating. The ages of the unknown fossils can be estimated by Equation 3, based on the k_L value determined from this calibrated sample. If two calibrated samples are available, Equation 3 is rewritten as:

$$\ln\left[\frac{1 + (D/L)}{1 - K'(D/L)}\right]_{t_2} - \ln\left[\frac{1 + (D/L)}{1 - K'(D/L)}\right]_{t_1} = (1 + K')k_L(t_2 - t_1)_{t_2 > t_1} \qquad (4)$$

where t_1 and t_2 are the ages of the samples determined by the absolute method. Equation 4 gives more precise ages than Equation 3.

B. Sample Preparation
1. Sample
A sediment sample at Osaka South Port in Osaka Bay was taken, and divided into 36 levels to the depth of 71.55 m. The shells were carefully separated from soils by immersion in water. The amounts of shells contained in the sediments at depths greater than 10 m are extremely low. The ages of the sediments at 8.5 m and 28.5 m are 3000 ± 110 and 14,000 ± 550 B.P. (Before Present, A.D. 1950 is the standard), as established by the radiocarbon method. Ages at 51.0 m and 64.5 m are not yet determined. Table 5 shows depth, chronostratigraphy, quality, color, shell content, and available radiocarbon ages of each of these four levels.

2. Sample Cleaning
About 0.2 g of the fossil shells are required for one analysis. First, shells must be washed with ultrasonication in 0.05 N HCl for 40 sec, next with water, then ethanol, and finally Soxhlet extraction with diethyl ether for 6 hr.[60] The fossil shells are then dried under reduced pressure.

3. Hydrolysis
After the sample has been washed and dried, it is placed in a Teflon®-lined screw cap Pyrex® tube. About 50 mℓ of 6 N HCl is added to the sample, then purged with nitrogen gas for 30 min to remove dissolved oxygen. After hydrolysis for 20 hr at 110°C, the hydrolysates are filtered through a quartz-fiber filter (0.45 μm). The filtrate is evaporated to dryness in a rotary vacuum evaporator.

4. Ion Exchange Purification
a. Cation Exchange
Bio-Rad® AG 5OW-X8 is used as a cation-exchange resin. First, resin (10 mℓ) is placed in a glass column that is fitted with a quartz wool plug at both ends. The column is washed sequentially with 30 mℓ each of 4 N HCl, water, 2 N NaOH, and water at a rate of 5 mℓ/min. This washing cycle is performed at least twice. Then, 30 mℓ of 4 N HCl is passed through this column, and washed with water until neutral as verified with a pH meter.

Table 5
SEDIMENTS AT OSAKA SOUTH PORT

Depth (m)	Chronostratigraphy		Sediment			Shell content (g)	^{14}C Age (year B.P.)
	Age	Formation	Quality	Color	Quantity (g)		
8.5	Holocene	Umeda bed	Silty clay	Dark bluish-gray	410	10	3,000 ± 110
28.5	Pleistocene	Nanko bed	Clay	Dark bluish-gray, slightly brown	430	0.6	14,000 ± 550
51.0	Pleistocene	Itami formation	Silty clay	Dark bluish-gray	360	1.0	28,800 ± 600
64.5	Pleistocene	Itami formation	Medium-grained sand with gravel	Dark gluish-gray	400	0.5	31,100 ± 580

The hydrolysate dissolved in a small amount of water is passed through this column at a rate of 2 mℓ/min. After washing with about 30 mℓ of water, the amino acids are eluted with 3 N NH$_4$OH. The effluent is taken to dryness in a rotary evaporator at a temperature below 50°C.

b. Anion Exchange

Bio-Rad® AG 1-X8 is used as an anion exchange resin. This column (5 mℓ) is washed with the acid-water-base-water cycle at least twice. Acid and base used are 30 mℓ each of 4 N HCl and 2 N NaOH. The final water washing after treatment with NaOH is continued until the column is neutral.

The solid residue from the cation exchange column is dissolved in water, and passed through this anion exchange column at a rate of 1 mℓ/min. After washing with about 30 mℓ of water, the amino acids are eluted with 1 N CH$_3$COOH. The eluate is evaporated in a rotary evaporator.

C. Ages of the Sediments

Figure 17A and B present chromatograms of the $N(O)$-TFA isopropyl esters of amino acids in the shells at depths of 8.5 and 51.0 m, respectively. Of various enantiomeric peaks, D-enantiomers can be found in alanine, valine, leucine, proline, serine, phenylalanine, glutamic acid, ornithine, and lysine. These constituents are also found in the shells at depths of 28.5 and 64.5 m. Comparing these two chromatograms, the D:L ratio of each amino acid contained in the shells increases with the depth or age of the sediment. The extent of racemization of alanine, valine, phenylalanine, and glutamic acid correlates especially well, and in proportion, to depth or age. Aspartic acid racemization has often been used to date fossil bones, due to its appropriate rate of racemization (half-life: about 15,000 years at 20°C).[41] Here, the aspartic acid peak was too small to be utilized; however, alanine, valine, phenylalanine, and glutamic acid were present in large quantities, thus enabling the sediments at all the levels attempted to be dated with high precision. Table 6 shows the D:L ratios of alanine, valine, phenylalanine, and glutamic acid at all depths determined, the racemization rate constant of each amino acid, and ages of the sediments at 51.0 and 64.5 m. The racemization rates are seen to follow the order: valine < glutamic acid < phenylalanine < alanine. The order provides evidence that these samples were not contaminated.[57] The ages derived from valine vary somewhat from the other data, which is probably due to the slower rate of racemization. However, the other ages coincide closely. To summarize these results, sediments at Osaka South Port are dated at 27,000 to 28,000 B.P. at 51.0 m and 29,000 to 30,000 B.P. at 64.5 m.

VI. SUMMARY AND CONCLUSIONS

Amino acid enantiomeric mixtures are derivatized to various $N(O,S)$-perfluoroacyl alkyl esters with subsequent direct resolution by GC both on Chirasil-Val® glass capillary columns and Heliflex™ Chirasil-Val® capillary columns.

Chirasil-Val® glass capillary columns show high resolution efficiency but are unstable for high-temperature operation. On the contrary, Heliflex™ Chirasil-Val® capillary columns are not as highly efficient for resolving amino acid enantiomeric derivatives, but are stable for high-temperature use (up to 210°C) without appreciable bleeding.

From the results of the various amino acid enantiomeric derivatives studied on Chirasil-Val® capillary columns, complete resolution and separation of a mixture of amino acids could not be obtained with a single run.

If we desire to determine completely the protein amino acid enantiomers on Heliflex™ Chirasil-Val®, they must be derivatized into two types of enantiomeric mixtures, $N(O,S)$-PFP isopropyl esters and $N(O,S)$-TFA ethyl ester derivatives.

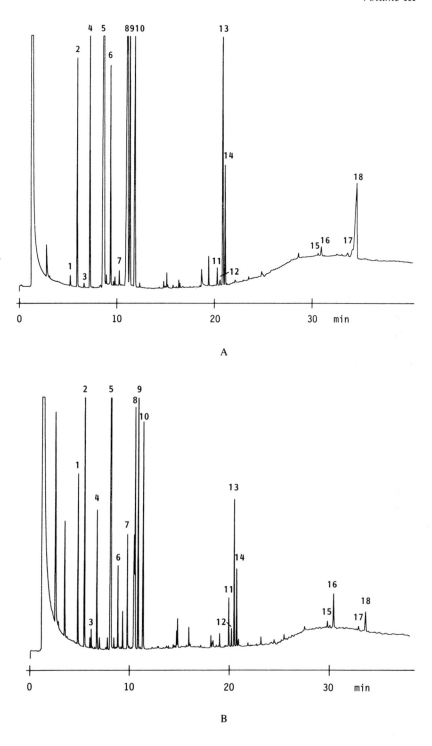

FIGURE 17. Gas chromatograms of TFA-isopropyl esters of amino acid in fossil shells at the depth of (A) 8.5 m and (B) 51.0 m. Column conditions: 25 m × 0.3 mm Chirasil-Val® capillary column; temperature: 90°C, 4-min hold and then programmed to 180°C at 4°/min. (1) D-Ala; (2) L-Ala; (3) D-Val; (4) L-Val; (5) Gly; (6) L-Ile; (7) D-Leu; (8) Pro; (9) L-Leu + D-Ser; (10) L-Ser; (11) D-Phe; (12) D-Glu; (13) L-Phe; (14) L-Glu; (15) D-Orn; (16) L-Orn; (17) D-Lys; (18) L-Lys.

Table 6
**RACEMIZATION OF AMINO ACIDS IN SHELLS AT
OSAKA SOUTH PORT**

Amino Acid	Depth (m)	D:L	Amino acid age (year B.P.)
Ala	8.5	0.063	$k = 1.62 \times 10^{-5} \text{year}^{-1}$
	28.5	0.237	
	51.0	0.432	27,700
	64.5	0.458	29,700
Val	8.5	0.023	$k = 5.36 \times 10^{-6} \text{year}^{-1}$
	28.5	0.082	
	51.0	0.139	24,900
	64.5	0.171	31,000
Phe	8.5	0.077	$k = 1.09 \times 10^{-5} \text{year}^{-1}$
	28.5	0.194	
	51.0	0.328	27,200
	64.5	0.350	29,400
Glu	8.5	0.087	$k = 7.19 \times 10^{-6} \text{year}^{-1}$
	28.5	0.164	
	51.0	0.255	27,300
	64.5	0.273	29,900

Also, the principle of a new method of dating fossils based on the racemization of amino acids and the practical working method of analysis are described in detail. Sediments at Osaka South Port along Osaka Bay are dated from the extent of racemization of alanine, valine, phenylalanine and glutamic acid in fossils shells from these sediments.

As this technique is dependent on the surrounding environment, one or more radiocarbon calibration samples are required in order to estimate ages accurately. Therefore, an absolute age should not be determined by this method, but has an advantage for samples that are outside the age and sample quantity requirements of the radiocarbon dating method.

Comparison of Chirasil-Val® glass capillary columns and Heliflex™ Chirasil-Val® fused silica columns has resulted in the following observations. The former has shown higher resolution (RN) than Heliflex™ Chirasil-Val®, but with regard to thermal stability, Heliflex™ Chirasil-Val® is more stable than Chirasil-Val® glass columns. Heliflex™ Chirasil-Val® columns have longer column lives, as they deteriorate more slowly. Chirasil-Val® capillary columns have been widely used for the resolution of optical antipodes of not only amino acids, but also amines, amino alcohols, and hydroxycarboxylic acids.[37]

The ages of the sediments at Osaka South Port in Osaka Bay were determined, based on the racemization of amino acids as measured by GC using a Chirasil-Val® capillary column. The required amount of fossils contained in the sediments for one analysis was less than 0.2 g. It provided a good model for testing the potential of age-dating from D:L ratios of several amino acids. The availability of Chirasil-Val® will increase with developments in biochemistry.

VII. COMMENTS ON THE METHOD

1. Amino acid enantiomeric analysis has been applied primarily to the determination of D:L ratios, or percentages of D-enantiomer, of individual amino acids, and not for total quantitative analysis. These are fundamentally different approaches, and slight losses of constituents during formation of derivatives are not critical if only ratios are needed.
2. If highly accurate data are desired, correction factors must be applied to account for D-enantiomers which occur as a result of the analytical procedure. This is especially important in the case of aspartic acid.

3. Of course, contamination markedly lowers the D-enantiomeric ratio. Every glass-made apparatus must be cleaned by washing with doubly distilled water, after soaking in a neutral cleaning solution for 1 week. It is desirable that sample pretreatment, derivatization, and final GC analysis be carried out in separate adjoining laboratory rooms.

4. All organic solvents are purified by distilling once or twice from commercially available reagents in an all-glass apparatus before use. Especially, 6 N HCl and H_2O must be carefully distilled at least two times to completely remove impurities.

5. Sample precut and back-flushing systems on the injection port of the gas chromatograph prevent less volatile impurities from entering the capillary column and effectively lengthen the column-life.

6. Deterioration and slow racemization of Chirasil-Val® could be observed after exceedingly high-temperature operation for long periods. It is preferable to use Chirasil-Val® below 210°C, with only short periods of operation to 240°C.

7. HCl/Alcohol should be protected from moisture, and should not be prepared in large quantity.

8. The fossils must be washed in dilute HCl (below 0.1 N) with ultrasonication to remove contaminants. Concentrated HCl causes effusion of bound amino acids, peptides, and proteins. Soxhlet extraction using diethyl ether as a solvent should be carried out to remove organic contaminants of human origin which greatly affect the determination of D:L ratios of amino acids.

9. After the ion exchange procedures, the eluates should be evaporated to dryness, using a rotary vacuum evaporator, at low temperature and as rapidly as possible to avoid racemization of amino acids. Similarly, if large volumes of solvent and reagents are used in the derivatization process, they should be removed by rotary evaporation rather than under a nitrogen stream with heating.

10. Cysteine, methionine, and tryptophan are reported to be very sensitive to oxygen. Cysteine can easily dimerize to cystine, methionine is oxidized to the sulfoxide, and tryptophan is decomposed with heating. Great care must be used during hydrolysis of samples, derivatization, and GC analysis to exclude oxygen as much as possible.

11. Before hydrolysis of the fossils, the 6 N HCl must be bubbled with nitrogen for at least 30 min in an ice-water bath to remove dissolved oxygen. If below nanogram amounts of amino acids are present in fossil samples, hydrolysis should be carried out using pressure-stable thick-walled glass. The tube should be sealed during immersion in liquid nitrogen under vacuum.

12. Other modified types of Chirasil-Val® phases are said to lack reproducible chemical properties, especially in regard to resolution factors. This is caused by discrepancies in optical activity that result from a serious shortcoming; the inability to further purify those modified silicone polymers.

ACKNOWLEDGMENT

This work has been generously supported by the Ministry of Education of Japan. We wish to thank Dr. M. Sakanoue and Dr. K. Megumi, for sharing samples with us.

REFERENCES

1. **Jaeger, H., Kloer, H. U., Ditschuneit, H., and Frank, H.,** Glass capillary gas-liquid chromatography of amino acids, in *Applications of Glass Capillary Gas Chromatography,* Chromatographic Science Ser., Vol. 15, Jennings, W. G., Ed., Marcel Dekker, New York, 1981, 331.

2. **Abe, I., Izumi, K., Kuramoto, S., and Musha, S.,** GC resolution of various D,L-amino acid derivatives on a Chirasil-Val capillary column, *HRC&CC,* 4, 549, 1981.
3. **Gil-Av, E., Charles-Sigler, R., and Fischer, G.,** Resolution of amino acids by gas chromatography, *J. Chromatogr.,* 17, 408, 1965.
4. **Pollock, G. E., Oyama, V. I., and Johnson, R. D.,** Resolution of racemic amino acids by gas chromatography, *J. Gas Chromatogr.,* 3, 174, 1965.
5. **Gil-Av, E., Charles-Sigler, R., Fischer, G., and Nurok, D.,** Resolution of optical isomers by gas liquid partition chromatography, *J. Gas Chromatogr.,* 4, 51, 1966.
6. **Pollock, G. E. and Oyama, V. I.,** Resolution and separation of racemic amino acids by gas chromatography and the application to protein analysis, *J. Gas Chromatogr.,* 4, 126, 1966.
7. **Pollock, G. E. and Kawauchi, A. H.,** Resolution of racemic aspartic acid, tryptophan, hydroxy and sulfhydryl amino acids by gas chromatography, *Anal. Chem.,* 40, 1356, 1968.
8. **Raulin, F. and Khare, B. N.,** Gas-liquid chromatographic resolution of several protein amino acid enantiomers on a packed column, *J. Chromatogr.,* 75, 13, 1973.
9. **Weygand, F., Prox, A., Schmidhammer, L., and König, W.,** Gas chromatographic investigation of racemization in peptide syntheses, *Angew. Chem.,* 75, 282, 1963.
10. **Halpern, B. and Westley, J. W.,** Optical resolution of DL-amino acids by gas chromatography and mass spectrometry, *Biochem. Biophys. Res. Commun.,* 19, 361, 1965.
11. **Halpern, B. and Westley, J. W.,** High sensitivity optical resolution of poly-functional amino acids by gas liquid chromatography, *Tetrahedron Lett.,* 21, 2283, 1966.
12. **Halpern, B. and Westley, J. W.,** Demonstration of the stereospecific action of microorganisms in soil by gas liquid chromatography, *Anal. Biochem.,* 17, 179, 1966.
13. **Dabrowiak, J. C. and Cooke, D. W.,** Gas-liquid chromatography of the optical isomers of threonine and allothreonine, *Anal. Chem.,* 43, 791, 1971.
14. **Bonner, W. A.,** The adaptation of diastereoisomeric S-prolyl dipeptide derivatives to the quantitative estimation of R- and S-leucine enantiomers, *J. Chromatogr. Sci.,* 10, 159, 1972.
15. **Bonner, W. A.,** Enantiomeric markers in the quantitative gas chromatographic analysis of optical isomers. Application to the estimation of amino acid enantiomers, *J. Chromatogr. Sci.,* 11, 101, 1973.
16. **Iwase, H. and Murai, A.,** Resolution of racemic amino acids by gas chromatography. II. N-Perfluoroacyl-L-prolyl derivatives, *Chem. Pharm. Bull.,* 22, 8, 1974.
17. **Gil-Av, E., Feibush, B., and Charles-Sigler, B.,** Separation of enantiomers by gas liquid chromatography with an optically active stationary phase, *Tetrahedron Lett.,* 10, 1009, 1966.
18. **Gil-Av, E. and Feibush, B.,** Resolution of enantiomers by gas liquid chromatography with optically active stationary phases. Separation on packed columns, *Tetrahedron Lett.,* 35, 3345, 1967.
19. **König, W. A., Parr, W., Lichtenstein, H. A., Bayer, E., and Oro, J.,** Gas chromatographic separation of amino acids and their enantiomers: nonpolar stationary phases and a new optically active phase, *J. Chromatogr. Sci.,* 8, 183, 1970.
20. **Feibush, B. and Gil-Av, E.,** Interaction between asymmetric solutes and solvents. Peptide derivatives as stationary phases in gas liquid partition chromatography, *Tetrahedron,* 26, 1361, 1970.
21. **Parr, W., Yang, C., Bayer, E., and Gil-Av, E.,** Interaction between asymmetric solutes and solvents: N-trifluoroacetyl-L-phenylalanyl-L-leucine cyclohexyl ester as solvent, *J. Chromatogr. Sci.,* 8, 591, 1970.
22. **Corbin, J. A., Rhoad, J. E., and Rogers, L. B.,** Effects of structure of peptide stationary phases on gas chromatographic separations of amino acid enantiomers, *Anal. Chem.,* 43, 327, 1971.
23. **Beitler, U. and Feibush, B.,** Interaction between asymmetric solutes and solvents. Diamides derived from L-valine as stationary phases in gas-liquid partition chromatography, *J. Chromatogr.,* 123, 149, 1976.
24. **Parr, W., Pleterski, J., Yang, C., and Bayer, E.,** Resolution of racemic amino acids by gas chromatography on optically active stationary phases, *J. Chromatogr. Sci.,* 9, 141, 1971.
25. **Feibush, B.,** Interaction between asymmetric solutes and solvents. N-Lauroyl-L-valyl-t-butylamide as stationary phase in gas liquid partition chromatography, *Chem. Commun.,* 544, 1971.
26. **Parr, W. and Howard, P.,** Separation of amino acid enantiomers by gas chromatography with an optically active stationary phase(N-TFA-L-valyl-L-leucine cyclohexyl ester), *Chromatographia,* 4, 162, 1971.
27. **Parr, W. and Howard, P. Y.,** Molecular interactions in a unique solvent-solute system, *J. Chromatogr.,* 71, 193, 1972.
28. **Grohman, K. and Parr, W.,** Investigation of the diastereoisomeric association complex for the separation of amino acid enantiomers on optically active stationary phases, *Chromatographia,* 5, 18, 1972.
29. **Andrawes, F., Brazell, R., Parr, W., and Zlatkis, A.,** Methionine dipeptide stationary phases for the resolution of enantiomers, *J. Chromatogr.,* 112, 197, 1975.
30. **König, W. A. and Nicholson, G. J.,** Glass capillaries for fast gas chromatographic separation of amino acid enantiomers, *Anal. Chem.,* 47, 951, 1975.
31. **Abe, I., Kohno, T., and Musha, S.,** Resolution of amino acids on optically active stationary phase by gas chromatography, *Chromatographia,* 11, 393, 1978.
32. **Feibush, B. and Gil-Av, E.,** Gas chromatography with optically active stationary phases, *J. Gas Chromatogr.,* 5, 257, 1967.

33. **Charles, R., Beitler, U., Feibush, B., and Gil-Av, E.,** Separation of enantiomers on packed columns containing optically active diamide phases, *J. Chromatogr.,* 112, 121, 1975.

34. **Charles, R. and Gil-Av, E.,** N-Docosanoyl-L-valyl-2-(2-methyl)-*n*-heptadecylamide as a stationary phase for the resolution of optical isomers in gas-liquid chromatography, *J. Chromatogr.,* 195, 317, 1980.

35. **Frank, H., Nicholson, G. J., and Bayer, E.,** Rapid gas chromatographic separation of amino acid enantiomers with a novel chiral stationary phase, *J. Chromatogr. Sci.,* 15, 174, 1977.

36. **Nicholson, G. J., Frank, H., and Bayer, E.,** Glass capillary gas chromatography of amino acid enantiomers, *HRC&CC,* 2, 411, 1979.

37. **Frank, H., Nicholson, G. J., and Bayer, E.,** Gas chromatographic-mass spectrometric analysis of optically active metabolites and drugs on a novel chiral stationary phase, *J. Chromatogr.,* 146, 197, 1978.

38. **Wulff, D.,** Chirasil-Val®, improved optically active stationary phases for gas chromatography, Pittsburgh Conf. Analytical and Applied Spectroscopy Abstracts, Atlantic City, N.J., 1982, 791.

39. **Hare, P. E. and Abelson, P. H.,** Racemization of amino acids in fossil shells, *Carnegie Inst. Washington Year.,* 66, 526, 1968.

40. **Hare, P. E. and Mitterer, R. M.,** Laboratory simulation of amino acid diagenesis in fossils, *Carnegie Inst. Washington Year.,* 67, 205, 1969.

41. **Hoopes, E. A., Peltzer, E. T., and Bada, J. L.,** Determination of amino acid enantiomeric ratios by gas liquid chromatography of N-trifluoroacetyl-L-prolyl-peptide methyl esters, *J. Chromatogr. Sci.,* 16, 556, 1978.

42. **Bada, J. L.,** The dating of fossil bones using the racemization of isoleucine, *Earth Planet. Sci. Lett.,* 15, 223, 1972.

43. **Matsu'ura, S. and Ueta, N.,** Fraction dependent variation of aspartic acid racemization age of fossil bone, *Nature (London),* 286, 883, 1980.

44. **Bada, J. L., Kvenvolden, K. A., and Peterson, E.,** Racemization of amino acids in bones, *Nature (London),* 245, 308, 1973.

45. **Bada, J. L.,** Racemization of amino acids in fossil bones and teeth from the Olduvai Gorge region, Tanzania, East Africa, *Earth Planet. Sci. Lett.,* 55, 292, 1981.

46. **Abe, I., Izumi, K., Kuramoto, S., and Musha, S.,** Age-dating of the rings of Yakusugi *(Cryptomeria japonica)* based on the racemization of amino acids, *Bunseki Kagaku(Jpn. Anal.),* 31, 427, 1982.

47. **Kvenvolden, K. A., Peterson, E., and Brown, F. S.,** Racemization of amino acids in sediments from Saanich Inlet, British Columbia, *Science,* 169, 1079, 1970.

48. **Bada, J. L. and Schroeder, R. A.,** Racemization of isoleucine in calcareous marine sediments: kinetics and mechanism, *Earth Planet. Sci. Lett.,* 15, 1, 1972.

49. **Mitterer, R. M.,** Ages and diagenetic temperatures of Pleistocene deposits of Florida based on isoleucine epimerization in Mercenaria, *Earth Planet. Sci. Lett.,* 28, 275, 1975.

50. **Bada, J. L. and Shou, M.-Y.,** Kinetics and mechanism of amino acid racemization in aqueous solution and in bones, in *Biogeochemistry of Amino Acids,* Hare, P. E., Hoering, T. C., and King, K., Jr., Eds., John Wiley & Sons, New York, 1980, 235.

51. **Liu, J. H. and Ku, W. W.,** Determination of enantiomeric N-trifluoroacetyl-L-prolyl chloride amphetamine derivatives by capillary gas chromatography/mass spectrometry with chiral and achiral stationary phases, *Anal. Chem.,* 53, 2180, 1981.

52. **Frank, H., Rettenmeiner, A., Weicker, H., Nicholson, G. J., and Bayer, E.,** Determination of enantiomer-labeled amino acids in small volumes of blood by gas chromatography, *Anal. Chem.,* 54, 715, 1982.

53. **Maeda, Y.,** The sea level changes of Osaka Bay from 12,000 B.P. to 6,000 B.P., *Koukogaku Shizenkagaku(Archaeol. Natural Sci.),* 9, 31, 1976.

54. **Smith, G. G. and Evans, R. C.,** The effect of structure and conditions on the rate of racemization of free and bound amino acids, in *Biogeochemistry of Amino Acids,* Hare, P. E., Hoering, T. C., and King, K., Jr., Eds., John Wiley & Sons, New York, 1980, 257.

55. **Bada, J. L. and Schroeder, R. A.,** Amino acid racemization reactions and their geochemical implications, *Naturwissenschaften,* 62, 71, 1975.

56. **Dungworth, G.,** Optical configuration and the racemization of amino acids in sediments and in fossils — a review, *Chem. Geol.,* 17, 135, 1976.

57. **Masters, P. M. and Bada, J. L.,** Amino acid racemization dating of bone and shell, in *Archaeological Chemistry II,* Gould, R. F., Ed., American Chemical Society, Chicago, 1978, chap. 8.

58. **Bada, J. L.,** Kinetics of racemization of amino acids as a function of pH, *J. Am. Chem. Soc.,* 95, 1371, 1972.

59. **Smith, G. G., Williams, K. M., and Wonnacott, D. M.,** Factors affecting the rate of racemization of amino acids and their significance to geochronology, *J. Org. Chem.,* 43, 1, 1978.

60. **Schroeder, R. A. and Bada, J. L.,** A review of the geochemical applications of the amino acid racemization reaction, *Earth Sci. Rev.,* 12, 1976.

Chapter 3

RESEARCH ON THE USE OF *s*-TRIAZINE DERIVATIVES FOR GC SEPARATION OF ENANTIOMERS OF AMINO ACIDS AND RELATED COMPOUNDS

Naobumi Ôi

TABLE OF CONTENTS

I. INTRODUCTION

In 1966, Gil-Av et al.[1] succeeded for the first time in separating *N*-trifluoroacetyl (TFA)-amino acid ester enantiomers by gas chromatography (GC) using glass capillary columns coated with *N*-TFA-L-isoleucine lauryl ester. The separation factors were rather small, but the achievement of resolution of enantiomers was one of the most striking demonstrations of the efficiency of GC.

Since this pioneering work, a number of different chiral phases have been prepared and evaluated. The three basic structural types of the stationary phases for GC separation of amino acid enantiomers are (1) *N*-acyl amino acid esters, (2) *N*-acyl peptide esters, and (3) *N*-acyl amino acid amides. These phases generally show good enantioselectivity for amino acids, but the drawback is their low temperature stability. Therefore, many studies were concerned with the influence of structural modifications on thermal properties of the stationary phase.

We have prepared *s*-triazine derivatives of amino acid esters, peptide esters and amino acid amides, and examined the chromatographic properties as chiral stationary phases for the separation of amino acid enantiomers.

II. *s*-TRIAZINE DERIVATIVES FOR SEPARATION OF AMINO ACID ENANTIOMERS

A. Derivatives of Amino Acid Esters

N-Acyl amino acid ester phases show enantioselectivity but suffer from column bleeding. For example, the working temperature of *N*-TFA-L-isoleucine lauryl ester was limited to about 110°C.[1]

The *s*-triazine derivatives of amino acid esters, such as *N*,*N'*,*N''*-[2,4,6-(1,3,5-triazine)triyl]-tris-L-valine isopropyl ester (OA-100; Structure 1), exhibited excellent properties for the separation of amino acid enantiomers as shown in Figure 1, and it was noted OA-100 could be used between 70 to 150°C.[2] This fact suggests the chemical structures containing the *s*-triazine ring are very effective for thermal stability.

Structure I

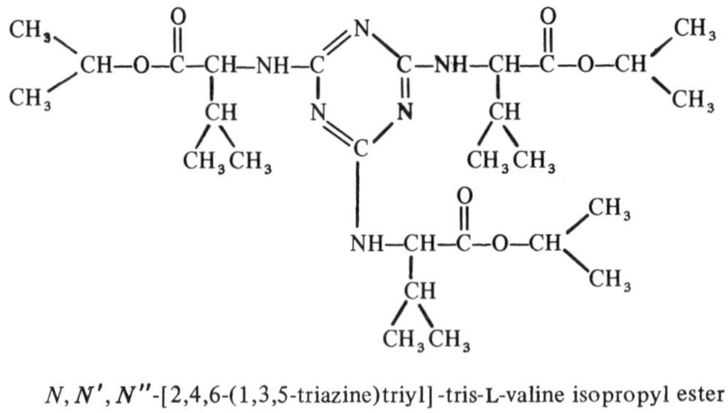

N, N', N''-[2,4,6-(1,3,5-triazine)triyl]-tris-L-valine isopropyl ester

(OA-100)

The separation of D,L-isomers of *N*-acyl amino acid esters by phases of the type $R_1CH(NHCOCF_3)COOR_2$ has been explained by the formation of hydrogen-bonded diastereomeric association complexes.[3] It is noted that OA-100, which has no —CONH— groups, gives essentially the same separation as *N*-acyl amino acid ester phases, and this result

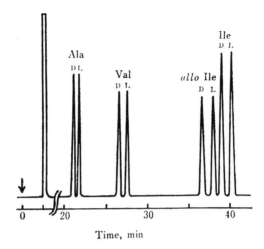

FIGURE 1. GC of *N*-TFA isopropyl esters of D,L-alanine, D,L-valine, D,L-*allo*-isoleucine, and D,L-isoleucine. Column: glass capillary (60 m × 0.25 mm I.D.) coated with OA-100; Temperature: 100°C. (From Ôi, N. et al., *Bunseki Kagaku (Jpn. Anal.),* 27, 637, 1978. With permission.)

suggests $-N = C -NH-$ groups attached to asymmetric carbon atoms make a contribution to
the separation of amino acid enantiomers.

B. Derivatives of Peptide Esters

As the inspection of models indicated a possible association complex of an *N*-TFA-amino acid ester with a dipeptide, containing three hydrogen bonds between NH and CO groups, higher separation factors were expected for dipeptide ester phases than those of amino acid ester phases.[3]

Indeed, the first dipeptide ester phase *N*-TFA-L-valyl-L-valine cyclohexyl ester produced separation factors so large as to permit resolution of amino acid enantiomers on a packed column.[4] However, this phase could be used only between 100 and 110°C, so new phases like *N*-TFA-L-phenylalanyl-L-phenylalanine cyclohexyl ester were synthesized in order to overcome the insufficient thermal stability.[5] The maximum permissible operating temperature of this phase was not higher than 160°C.

s-Triazine derivatives of dipeptide esters, such as *N,N'*-[2,4-(6-ethoxy-1,3,5-triazine)diyl]-bis-(L-valyl-L-valine isopropyl ester) (OA-200; Structure 2),[6] showed good enantioselectivity for amino acids as shown in Figure 2. The fact that OA-200 can be used between 80 and 170°C shows again incorporation of the *s*-triazine ring is effective for the thermal stability of stationary phases.

Structure II

N, N'-[2,4-(6-ethoxy-1,3,5-triazine)diyl] -bis-(L-valyl-L-valine isopropyl ester)

(OA-200)

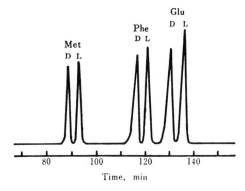

FIGURE 2.   GC of *N*-TFA isopropyl esters of D,L-methionine, D,L-phenylalanine, and D,L-glutamic acid. Column: glass capillary (30 m × 0.25 mm I.D.) coated with OA-200; temperature: 110°C. (From Ôi, N. et al., *Bunseki Kagaku (Jpn. Anal.)*, 28, 69, 1979. With permission.)

Table 1
SEPARATION FACTORS OF
N-PENTAFLUOROPROPIONYL-D,L-
AMINO ACID ISOPROPYL ESTERS

Compound	Column temp. (°C)	Chiral stationary phase	
		OA-200	OA-300
Met	140	1.035	1.087
Glu	140	1.028	1.067
Phe	140	1.030	1.086
Lys	170	1.021(7)	1.069(7)

Feibush and Gil-Av[3] attempted to investigate the effect of extending the peptide chain by synthesizing *N*-TFA-L-valyl-L-valyl-L-valine isopropyl ester. As this compound has a melting point as high as 202°C, they studied the chromatographic properties of the lower-melting binary mixtures of this compound and *N*-TFA-L-valyl-L-valine isopropyl ester. It was seen that the addition of the *N*-TFA-tripeptide ester tended to depress slightly the resolution of *N*-TFA-D,L-alanine *tert*-butyl ester.

Ôi et al.[7] prepared an *s*-triazine derivative of a tripeptide ester *N,N'*-[2,4-(6-ethoxy-1,3,5-triazine)diyl]-bis-(L-valyl-L-valyl-L-valine isopropyl ester) (OA-300; Structure 3). It is interesting that a result of the incorporation with a *s*-triazine ring is a decrease in the melting point. OA-300 has a melting point as high as 136°C, and can be used without the aid of another low-melting stationary phase. It is emphasized that this new phase shows even better properties for enantiomer separation of amino acids than the corresponding *s*-triazine derivative of the dipeptide ester, OA-200. Table 1 shows the comparison of separation factors for amino acid enantiomers on these two phases. The efficiency of the longer peptide chain is clearly seen in the behavior for the separation of *N*-acyl-D,L-amino acid esters. Moreover, OA-300 has much better thermal stability than OA-200. Thermogravimetric analysis of OA-300 showed that bleeding should start only at about 200°C. A stable baseline was obtained when operating at 180°C with the instrument set at 4×10^{-11} full-scale deflection as can be seen in Figure 3.

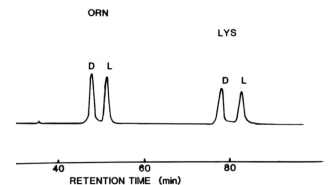

FIGURE 3. GC of *N*-pentafluoropropionyl isopropyl esters of D,L-orni-
thine and D,L-lysine. Column: glass capillary (40 m × 0.25 mm I.D.)
coated with OA-300; temperature: 180°C.

Structure III

CH₃ ... O—C—CH—NH—C—CH—NH—C—CH—NH—C ... C—NH—CH—C—NH—CH—C—NH—CH—C—O—CH ... CH₃

N,N′-[2,4-(6-ethoxy-1,3,5-triazine)diyl]-bis-(L-valyl-L-valyl-L-valine isopropyl ester)

(OA-300)

C. Derivatives of Amino Acid Amides

It is well known that the highest efficiency for amino acid enantiomer separation is obtained
on amino acid amide phases with the structure RCONHCH(*i*-pr)CONHR′ where R is un-
branched and R′ is a tertiary group.[8,9]

Structure IV

$(CH_3)_3C-NH-C-CH-(CH_2)_4-NH-C$ ... $C-NH-(CH_2)_4-CH-C-NH-C(CH_3)_3$

$CH_3(CH_2)_{10}-C-NH$ ... $NH-C-(CH_2)_{10}CH_3$

$NH-(CH_2)_4-CH-C-NH-C(CH_3)_3$

$NH-C-(CH_2)_{10}CH_3$

N,N′,N″-[2,4,6-(1,3,5-triazine)triyl]-tris-(*N*α-lauroyl-L-lysine-tert-butylamide)

(OA-400)

Ôi et al.[10] synthesized an *s*-triazine derivative of an amino acid amide, *N,N′,N″*-[2,4,6-
(1,3,5-triazine)triyl]-tris-(*N*α-lauroyl-L-lysine-*tert*-butylamide) (OA-400; Structure 4).[10] This
phase has even better enantioselectivity than that of OA-300, and high separation factors

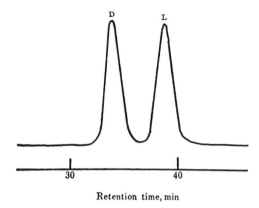

FIGURE 4.   GC of *N*-TFA-D,L-leucine isopropyl ester.
Column: 2.8 m × 0.5 mm I.D., containing Shimalite®
W (170/200 mesh) coated with 10 w/w% OA-400; tem-
perature: 130°C. (From Ôi, N. et al., *Bunseki Kagaku
(Jpn. Anal.)*, 29, 270, 1980. With permission.)

permit the separation of some amino acid enantiomers on packed columns as shown in Figure
4. Unfortunately, the working temperature of this phase was limited to about 150°C.

III. SEPARATION OF ENANTIOMERS OF RELATED COMPOUNDS

It should be emphasized that *s*-triazine derivatives of amino acid esters, dipeptide esters,
tripeptide esters, and amino acid amides have excellent properties in relation to enantiomer
separation of not only amino acids but also related compounds, such as dipeptides, amines,
carboxylic acids, etc.

A. Dipeptides

Hitherto, the GC separation of diastereomeric dipeptides has been achieved with ordinary
optically inactive stationary phases, and the enantiomers of dipeptides were not resolved.
Therefore it was necessary to hydrolyze the dipeptide and determine the configuration of
the resulting amino acids in order to distinguish D-L from L-D or D-D from L-L diastereoisomers.

We have succeeded in the direct resolution of the enantiomers of some linear and cyclic
dipeptides by use of OA-300.[11] As shown in Figure 5, the L-D, D-L, D-D, and L-L isomers
of *N*-TFA-D,L-alanyl-D,L-alanine isopropyl ester were resolved. Cyclic alanylalanine (2,5-
dimethyldiketopiperazine) was resolved into three peaks as shown in Figure 6. As the D-L
and L-D isomers are identical in this dipeptide, their peaks are superimposed and the ratio
of the three peak areas is 1:1:2.

B. Amines

On *N*-TFA-amino acid ester phases or *N*-TFA-peptide ester phases, enantiomeric amines
could not be separated efficiently, and their resolution was best achieved with stationary
phases such as carbonyl-*bis*-(L-valine isopropyl ester) and *N*-lauroyl-(*S*)-α-(1-naph-
thyl)ethylamine.[12,13] However, working temperatures were limited to about 120 and 150°C,
respectively, and the retention times of aromatic amines were very long.

Ôi et al.[2,6,7,10] have found *N*-acyl derivatives of amines were resolved with *s*-triazine
derivatives as chiral stationary phases.

Especially, some arylalkylamine enantiomers could be well separated on OA-300 with
short retention times as shown in Table 2.[14] Recently, Charles and Gil-Av[15] reported a
thermostable diamide phase *N*-docosanoyl-L-valine-2-(2-methyl)-*n*-heptadecylamide [I]

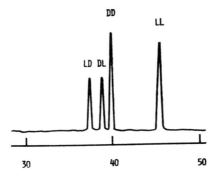

FIGURE 5. GC of *N*-TFA-D,L-alanyl-D,L-alanine isopropyl ester. Column: glass capillary (40 m × 0.25 mm I.D.) coated with OA-300; temperature: 180°C. (From Ôi, N. et al., *J. Chromatogr.*, 202, 302, 1980. With permission.)

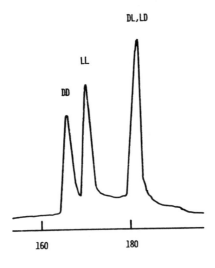

FIGURE 6. GC of cyclo-D,L-alanyl-D,L-alanine. Column: glass capillary (40 m × 0.25 mm I.D.) coated with OA-300; temperature: 185°C. (From Ôi, N. et al., *J. Chromatogr.*, 202, 302, 1980. With permission.)

showed stereoselectivity for *N*-TFA-amines, but comparison with data for OA-300 [II] shows lower separation factors. (1-Phenylethylamine: $\alpha = 1.035$ on [I] (130°C), 1.062 on [II] (140°C); 1-(α-naphthyl)ethylamine (180°C): $\alpha = 1.033$ on [I], 1.063 on [II].)

C. Carboxylic Acids

The enantiomers of α-branched carboxylic acids which do not possess a NH group directly linked to the asymmetric carbon atom were separated for the first time by Weinstein et al.[13]

Table 2
GC SEPARATION OF THE ENANTIOMERS OF ARYLALKYLAMINES AS N-ACYL DERIVATIVES

Compound[a]	Column temp. (°C)	Retention time (min)		Separation factor (α)	Resolution (R)
		Peak I (*R*-isomer)	Peak II (*S*-isomer)		
I α-Phenylethylamine	150	9.1	9.6	1.057	1.68
II α-Phenyl-*n*-propylamine	150	11.7	12.3	1.051	1.50
III α-(2,5-Xylyl)ethylamine	150	13.0	13.8	1.063	1.82
IV α-(1-Naphthyl)ethylamine	180	23.8	25.3	1.063	3.41
V α-(2-Naphthyl)ethylamine	180	32.4	34.2	1.056	3.46
VI α,β-Diphenylethylamine	180	47.8	48.9	1.023	1.34
VII α-Phenyl-β-(4-tolyl)ethylamine	180	40.0	41.0	1.025	1.28

[a]   I, II, III, VI: *N*-TFA-derivatives; IV, V, VII: *N*-PFP-derivatives. Separations on AO-300 stationary phase.

From Ôi, N., Horiba, M., and Kitahara, H., *Bunseki Kagaku (Jpn. Anal.)*, 28, 482, 1979. With permission.

Table 3
GC SEPARATION OF THE ENANTIOMERS OF α-ALKYLPHENYLACETIC ACID ISOPROPYLAMIDES

	Compound	Retention time (min)[a] Peak I (*S*-isomer)	Peak II (*R*-isomer)	Separation factor (α)	Resolution (R)
I	α-Methylphenylacetic acid	11.8	12.2	1.037	1.51
II	α-Ethylphenylacetic acid	15.5	16.0	1.033	1.28
III	α-Isopropylphenylacetic acid	17.0	17.6	1.035	1.33
IV	α-Isopropyl-4-methyl-phenyl-acetic acid	23.3	24.2	1.039	1.37
V	α-Isopropyl-4-*tert*-butyl-phenyl-acetic acid	44.9	47.0	1.047	1.98
VI	α-Isopropyl-4-chloro-phenyl-acetic acid	48.9	50.6	1.034	1.84
VII	α-Isopropyl-4-methoxy-phenyl-acetic acid	58.2	60.1	1.033	1.46
VIII	α-Isopropyl-4-bromo-phenyl-acetic acid[b]	47.6	48.9	1.027	1.38

[a]   Column temperature, 180°C. Separations on AO-300 stationary phase.
[b]   Resolved in the form of *tert*-butylamide.

From Ôi, N., Horiba, M., and Kitahara, H., *Bunseki Kagaku (Jpn. Anal.)*, 28, 607, 1979. With permission.

with *N*-acyl derivatives of chiral amines as the stationary phase. Relatively large separation factors were obtained for the aromatic acids, but retention times were very long, as the column temperatures with these phases were limited to 150°C.

We have succeeded in the excellent separation of many α-phenylcarboxylic acid enantiomers in the form of isopropyl or *tert*-butylamides on OA-300 as shown in Table 3.[16] The rather high column temperature serves for rapid elution. For example, the racemic α-methylphenylacetic acid isopropylamide was separated within 15 min with a separation factor 1.037 at 180°C.

D. α-Hydroxycarboxylic Acid Esters

α-Amino acid enantiomers can be easily resolved in the form of their *N*-acyl ester derivatives by GC with optically active stationary phases. However, α-hydroxycarboxylic acid enantiomers in the form of *O*-acyl esters have never been separated, and it has been necessary to use the corresponding *O*-acylamide derivatives to resolve the enantiomers. This has been considered to be due to the absence of a nitrogen-attached hydrogen in *O*-acyl α-hydroxycarboxylic acid esters.

Ôi et al.[17] found enantiomers of several α-hydroxycarboxylic acid esters were resolved into their antipodes with *s*-triazine derivatives (OA-200, OA-300, OA-400) when the α-hydroxy group was not acylated. An example chromatogram is shown in Figure 7. The fact that enantiomers could not be resolved under the same chromatographic conditions when the α-hydroxy group was acylated indicates the free α-hydroxy group makes a large contribution to the separation of α-hydroxycarboxylic acid ester enantiomers.

E. Alcohols

Karagounis and Lippold[18] claimed the separation of racemic 2-butanol on diethyl *d*-tartrate, but their results could not be reproduced by Goldberg and Ross.[19] Berrod et al.[20] studied the behavior of chiral alcohols by GC on chiral phases, and found the separation was insufficient to observe graphically the commencement of resolution.

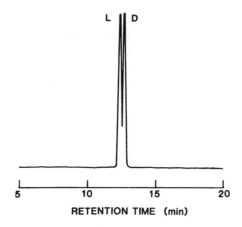

FIGURE 7.   GC of D,L-*n*-hexyl lactate. Column: glass capillary (40 m × 0.25 mm I.D.) coated with OA-300 + OA-200 (1:1); temperature: 130°C. (From Ôi, N. et al., *J. Chromatogr.*, 206, 143, 1981. With permission.)

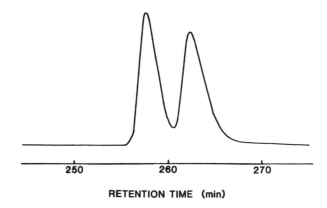

FIGURE 8.   GC of racemic 1-phenyl-2,2,2-trifluoroethanol. Column: glass capillary (95 m × 0.25 mm I.D.) coated with OA-400; temperature: 100°C. (From Ôi, N. et al., *J. Chromatogr.*, 208, 404, 1981. With permission.)

Recently, Ôi et al.[21,22] accomplished the first direct separation of some chiral alcohols by GC on *s*-triazine derivatives of peptide esters and amino acid amides (OA-200, OA-300, OA-400) as optically active stationary phases.

Figure 8 shows the gas chromatogram of racemic 1-phenyl-2,2,2-trifluoroethanol. However, in some chiral alcohols separation factors are rather small and longer retention times were required for efficient separation. Horiba et al.[23] succeeded in separating such alcohol enantiomers in the form of *N*-isopropyl urethane derivatives on OA-300.

F. Organophosphoroamidothioates

Direct separation of enantiomers of organophosphoroamidothioates containing an asymmetric phosphorous atom, such as racemic *O*-ethyl-*O*-(3-trifluoromethylphenyl)-*N*-isopropylphosphoroamidothioate has been accomplished with OA-300.[24,25] A typical chromatogram is shown in Figure 9.

It is noted that OA-300 is well suited to the separation of enantiomers of these organo-

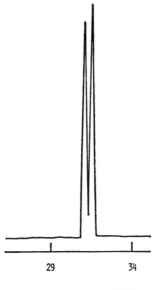

29   34

RETENTION TIME (MIN)

FIGURE 9.   GC of racemic *O*-ethyl-
O-(3-trifluoromethylphenyl) *N*-iso-
propylphosphoroamidothioate. Column: glass capillary (60 m × 0.25
mm I.D.) coated with OA-300; temperature: 180°C. (From Ôi, N. et al.,
Agric. Biol. Chem., 43, 2403, 1979.
With permission.)

phosphoroamidothioates which have no asymmetric carbon atom, although OA-300 was synthesized primarily for the purpose of enantiomer separation of amino acids which contain carbon as the sole asymmetric atom.

The resolution should be attributed to the diastereoisomeric intermolecular interaction containing hydrogen bonding between   S   group and   O   group.

$$\underset{\text{–P–NH–}}{\overset{\parallel}{}}\qquad\underset{\text{–C–NH–}}{\overset{\parallel}{}}$$

IV. DETERMINATION OF OPTICAL ISOMERS

s-Triazine derivatives of amino acid esters, peptide esters, and amino acid amides have efficient enantioselectivity for amino acids. Each amino acid shows good peak resolution, respectively, and the optical purity can be easily determined. These phases also have good thermal stability, but working temperatures are unfortunately insufficient for the rapid analysis of all common protein amino acids in one run by the use of a wide temperature program.

It is to be noted that these *s*-triazine derivatives are efficient stationary phases for the separation of various enantiomers of related compounds such as dipeptides, amines, etc. Examples of the determination of optical isomers of amino acids, amines, carboxylic acids, and alcohols are described.

A. Proline

Proline is used as a chiral reagent in the form of the *N*-acyl chloride for conversion of various alcohol, amine, and amino acid enantiomers into diastereoisomers to resolve the

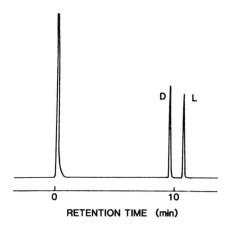

FIGURE 10.   GC of *N*-TFA-D,L-proline isopropylamide. Column: glass capillary (40 m × 0.25 mm I.D.) coated with OA-300; temperature: 180°C. (From Ôi, N. et al., *J. Chromatogr.*, 202, 299, 1980. With permission.)

optical isomers using optically inactive stationary phases; therefore, it is important to estimate the optical purity of proline.

As is well known, proline shows the lowest separation factor of all racemic amino acids in their *N*-acyl ester form, and this behavior is thought to be due to the secondary amide group which has no hydrogen left on its nitrogen atom after acylation. Therefore, the direct separation method has been considered to be insufficient for the analysis of proline enantiomers.

Recently Ôi et al.[26] found the enantiomers of proline could be separated with excellent separation factors in the form of *N*-acyl isopropylamides using *s*-triazine derivatives as stationary phases. Baseline separation was achieved within 15 min as shown in Figure 10. A small amount of D-isomer can be easily determined, and the optical isomer ratios determined by this method were in good agreement with those obtained by the diastereomeric determination method in which 1-menthol was used as a chiral reagent. When OA-400 is used as a chiral stationary phase, a high separation factor allows the separation with a packed column at a moderate temperature as shown in Figure 11.

1. Analytical Procedure[27]

To a solution of 100 mg of isopropylamine in 1 mℓ of toluene, 1 mℓ of *N*-TFA-prolyl chloride/toluene solution (50 to 70 mg/mℓ) was slowly added with stirring. The mixture was kept at room temperature for 10 min and acidified with 3 mℓ of 1 *N* HCl. After stirring, the organic phase was separated and dried over anhydrous sodium sulfate. A 1-μℓ volume of this solution was injected for GC analysis.

The chromatographic conditions used were as follows: column, 40 m × 0.25 mm I.D. glass capillary coated with OA-300; column temperature, 180°C; injector and detector temperature, 220°C; carrier gas, helium at a flow rate of 0.7 mℓ/min; split ratio, 1:100.

The peak areas were measured by using a digital integrator. The ratio of optical isomers was obtained from the ratio of the peak area of each isomer to the total peak area of the two isomers.

B. 1-Phenylethylamine

1-Phenylethylamine is used as a chiral reagent for resolution of enantiomeric pairs of various chiral acids.

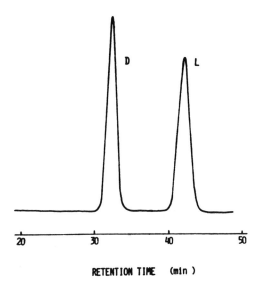

FIGURE 11.   GC of *N*-TFA-D,L-proline isopropylamide. Column: 2 m × 3 mm I.D., containing Chromosorb® W, AW, DMCS (100/120 mesh) coated with 5 w/w% OA-400; temperature: 130°C. (From Ôi, N. et al., *J. Chromatogr.*, 202, 299, 1980. With permission.)

The enantiomers of 1-phenylethylamine are separated in the form of their *N*-TFA derivatives on OA-200 as shown in Figure 12.

1. Analytical Procedure[28]

A mixture of 100 mg of 1-phenylethylamine and 3 mℓ of a solution of 2 mℓ of trifluoroacetic anhydride (TFAA) in toluene containing 10% ethyl acetate was kept at room temperature for 5 min, then 5 mℓ of water were added to decompose the excess of TFAA. The organic phase was separated and dried over anhydrous sodium sulfate and 1 μℓ of this solution was injected for GC analysis. The chromatographic conditions used were as follows: column, 30 m × 0.25 mm I.D. glass capillary coated with OA-200; column temperature, 150°C; injector and detector temperature, 220°C; carrier gas, helium at a flow rate of 0.7 mℓ/min; split ratio, 1:50.

The peak areas and ratio of optical isomers were determined as above.

C. 2-(4-Chlorophenyl)Isovaleric Acid

2-(4-Chlorophenyl)isovaleric acid is an acid moiety of a new synthetic pyrethroid, Fenvalerate, which has been developed as an agricultural pest control agent. The enantiomers of 2-(4-chlorophenyl)isovaleric acid are well separated in the form of their isopropylamide derivatives on OA-300 as shown in Figure 13.

1. Analytical Procedure[29]

To a solution of 15 mg of 2-(4-chlorophenyl)isovaleric acid and 20 mg of *N,N'*-dicyclohexyl carbodiimide in 0.5 mℓ of toluene, 1 mℓ of a solution of 40 mg of isopropylamine in toluene was added with stirring. The solution was kept at room temperature for 30 min. The mixture was acidified with 1 mℓ of 1 *N* HCl. Then, the organic phase was separated and dried over anhydrous sodium sulfate. A 3-μℓ volume of this solution was injected for GC analysis. The chromatographic conditions used were as follows: column, 40 m × 0.25 mm I.D. glass capillary coated with OA-300; column temperature, 180°C; injector and

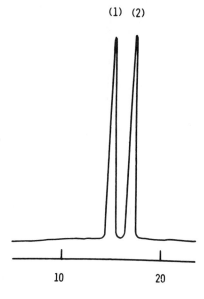

FIGURE 12. GC of racemic *N*-TFA-1-phenylethylamine. Column: glass capillary (30 m × 0.25 mm I.D.) coated with OA-200; temperature: 150°C. (1) *R*(+)-isomer; (2) *S*(−)-isomer. (From Horiba, M. et al., *Agric. Biol. Chem.*, 44, 2987, 1980. With permission.)

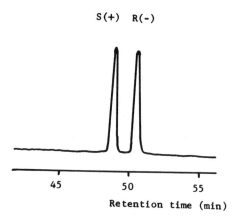

FIGURE 13. GC of racemic 2-(4-chlorophenyl)isovaleric acid isopropylamide. Column: glass capillary (40 m × 0.25 mm I.D.) coated with OA-300; temperature: 180°C.

detector temperature, 220°C, carrier gas, helium at a flow rate of 0.6 mℓ/min; split ratio, 1:100.

The peak areas and ratio of optical isomers were determined as above.

D. Allethrolone

Allethrolone (2-allyl-4-hydroxy-3-methyl-2-cyclopentene-1-one) is an important constit-

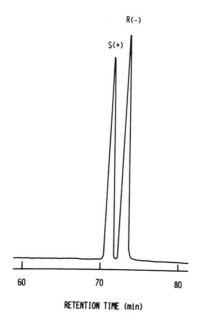

FIGURE 14. GC of isopropyl urethane derivative of racemic allethrolone. Column: glass capillary (42 m × 0.25 mm I.D.) coated with OA-300; temperature: 180°C. (From Horiba, M. et al., *Agric. Biol. Chem.*, 46, 281, 1982. With permission.)

uent of synthetic insecticidal pyrethroids. As is well known, optical isomers of pyrethroid esters have different toxicities for insects, depending on the configuration of the alcoholic group in the molecule, therefore it is important to establish the ratio of optical isomers.

Although the enantiomers of allethrolone could be directly separated on OA-300,[22] the separation factor was rather small and the retention time was very long. Recently, excellent separation of allethrolone enantiomers was obtained in the form of *N*-isopropylurethane with OA-300.[23] A typical chromatogram is shown in Figure 14.

1. Analytical Procedure[23]

A mixture of 20 mg of allethrolone, 0.1 mℓ of isopropyl isocyanate, and 0.1 mℓ of dry pyridine was heated on a water bath at 80 to 90°C for 1 hr. After cooling, 0.5 mℓ of methanol was added to decompose the excess isopropyl isocyanate. A 1 μℓ volume of this solution was injected for GC analysis. The chromatographic conditions used were as follows: column, 42 m × 0.25 mm I.D. glass capillary coated with OA-300; column temperature, 180°C; injector and detector temperature, 210°C; carrier gas, helium at a flow rate of 0.78 mℓ/min; split ratio, 1:80.

The ratio of optical isomers was determined as above.

V. CONCLUSIONS

For the separation of amino acid enantiomers by GC, three types of chiral stationary phases (amino acid esters, peptide esters, and amino acid amides) are generally efficient. In order to improve the thermal stability, many studies have been concerned with the influence on column operating temperature by modification of the stationary phase chemical structure.

As one of such studies, *s*-triazine derivatives of amino acid esters, peptide esters, and amino acid amides were prepared and their chromatographic properties investigated.

These new phases, which have relatively good enantioselectivity and thermal stability, are available for the analysis of amino acid enantiomers. It is especially noted that OA-300 (a tripeptide ester derivative) shows both high separation factors and high working temperatures. It should also be emphasized that these *s*-triazine derivatives have excellent enantioselectivity for not only amino acids, but also for dipeptides, amines, carboxylic acids, alcohols, etc.

We consider optical isomers of various compounds should be amenable to routine quantitative analysis with these novel phases.

REFERENCES

1. **Gil-Av, E., Feibush, B., and Charles-Sigler, R.,** Separation of enantiomers by gas liquid chromatography with an optically active stationary phase, *Tetrahedron Lett.,* 10, 1009, 1966.
2. **Ôi, N., Moriguchi, K., Matsuda, M., Shimada, H., and Hiroaki, O.,** *s*-Triazine derivatives of amino acid esters as novel stationary phases for the separation of amine and amino acid enantiomers by gas chromatography, *Bunseki Kagaku (Jpn. Anal.),* 27, 637, 1978.
3. **Feibush, B. and Gil-Av, E.,** Interaction between asymmetric solutes and solvents. Peptide derivatives as stationary phases in gas-liquid partition chromatography, *Tetrahedron,* 26, 1361, 1970.
4. **Gil-Av, E. and Feibush, B.,** Resolution of enantiomers by gas liquid chromatography with optically active stationary phases. Separation on packed columns, *Tetrahedron Lett.,* 35, 3345, 1967.
5. **König, W. A. and Nicholson, G. J.,** Glass capillaries for fast gas chromatographic separation of amino acid enantiomers, *Anal. Chem.,* 47, 951, 1975.
6. **Ôi, N., Takeda, H., Shimada, H., and Hiroaki, O.,** *s*-Triazine derivatives of dipeptide esters as novel stationary phases for the separation of amine and amino acid enantiomers by gas chromatography, *Bunseki Kagaku (Jpn. Anal.),* 28, 69, 1979.
7. **Ôi, N., Hiroaki, O., and Shimada, H.,** *s*-Triazine derivatives of tripeptide esters as novel stationary phases for the separation of amine and amino acid enantiomers by gas chromatography, *Bunseki Kagaku (Jpn. Anal.),* 28, 125, 1979.
8. **Feibush, B.,** Interaction between asymmetric solutes and solvents. N-Lauroyl-L-valyl-*t*-butylamide as stationary phase in gas-liquid partition chromatography, *Chem. Commun.,* 544, 1971.
9. **Beitler, U. and Feibush, B.,** Interaction between asymmetric solutes and solvents. Diamides derived from L-valine as stationary phase in gas-liquid partition chromatography, *J. Chromatogr.,* 123, 149, 1976.
10. **Ôi, N., Hiroaki, O., Shimada, H., Horiba, M., and Kitahara, H.,** A *s*-triazine derivative of L-lysine amide as a stationary phase for the separation of amino acid enantiomers by gas chromatography, *Bunseki Kagaku (Jpn. Anal.),* 29, 270, 1980.
11. **Ôi, N., Horiba, M., Kitahara, H., and Shimada, H.,** Gas chromatographic separation of enantiomers of some dipeptides on an optically active stationary phase, *J. Chromatogr.,* 202, 302, 1980.
12. **Feibush, B. and Gil-Av, E.,** Gas chromatography with optically active stationary phases. Resolution of primary amines, *J. Gas Chromatogr.,* 257, 1967.
13. **Weinstein, S., Feibush, B., and Gil-Av, E.,** N-Acyl derivatives of chiral amines as novel readily prepared phases for the separation of optical isomers by gas chromatography, *J. Chromatogr.,* 126, 97, 1976.
14. **Ôi, N., Horiba, M., and Kitahara, H.,** Gas chromatographic separation of arylalkylamine enantiomers with a chiral stationary phase, *Bunseki Kagaku (Jpn. Anal.),* 28, 482, 1979.
15. **Charles, R. and Gil-Av, E.,** N-Docosanoyl-L-valine-2-(2-methyl)-*n*-heptadecylamide as a stationary phase for the resolution of optical isomers in gas-liquid chromatography, *J. Chromatogr.,* 195, 317, 1980.
16. **Ôi, N., Horiba, M. and Kitahara, H.,** Gas chromatographic separation of α-alkylphenylacetic acid enantiomers with a chiral stationary phase, *Bunseki Kagaku (Jpn. Anal.),* 28, 607, 1979.
17. **Ôi, N., Kitahara, H., Horiba, M., and Doi, T.,** Gas chromatographic separation of α-hydroxycarboxylic acid ester enantiomers using amino acid derivatives as chiral stationary phase, *J. Chromatogr.,* 206, 143, 1981.
18. **Karagounis, G. and Lippold, G.,** Gaschromatographische spaltung racemischer verbindungen, *Naturwissenschaften,* 46, 145, 1959.
19. **Goldberg, G. and Ross, W. A.,** Separation of optical isomers by gas-liquid chromatography, *Chem. Ind. (London),* 657, 1962.

20. **Berrod, G., Bourdon, J., Dreux, J., Longeray, R., Moreau, M., and Schifter, P.,** Study of the behavior of chiral alcohols by gas chromatography on chiral phases, *Chromatographia,* 12, 150, 1979.
21. **Ôi, N., Doi, T., Kitahara, H., and Inda, Y.,** Direct separation of some alcohol enantiomers by gas chromatography with optically active stationary phases, *Bunseki Kagaku (Jpn. Anal.),* 30, 79, 1981.
22. **Ôi, N., Doi, T., Kitahara, H., and Inda, Y.,** Direct separation of some alcohol enantiomers by gas chromatography with amino acid derivatives as chiral stationary phases, *J. Chromatogr.,* 208, 404, 1981.
23. **Horiba, M., Kida, S., Yamamoto, S., and Ôi, N.,** Gas chromatographic determination of optical purity of allethrolone on a chiral stationary phase, *Agric. Biol. Chem.,* 46, 281, 1982.
24. **Ôi, N., Shimada, H., Hiroaki, O., Horiba, M., and Kitahara, H.,** Gas chromatographic resolution of enantiomers of organophosphoroamidothioate on an optically active stationary phase, *Bunseki Kagaku (Jpn. Anal.),* 28, 64, 1979.
25. **Ôi, N., Horiba, M., and Kitahara, H.,** Direct gas chromatographic resolution of organophosphoroamidothioate enantiomers on a chiral stationary phase, *Agric. Biol. Chem.,* 43, 2403, 1979.
26. **Ôi, N., Horiba, M., and Kitahara, H.,** Gas chromatographic separation of amino acid amide enantiomers on optically active stationary phases, *J. Chromatogr.,* 202, 299, 1980.
27. **Horiba, M., Kitahara, H., Yamamoto, S., and Ôi, N.,** Gas chromatographic determination of optical purity of a chiral derivatization reagent, N-acyl-L-prolyl chloride, *Agric. Biol. Chem.,* 44, 2989, 1980.
28. **Horiba, M., Kitahara, H., Yamamoto, S., and Ôi, N.,** Gas chromatographic determination of optical isomers of α-phenylethylamine and α-phenyl-β-(p-tolyl)ethylamine on optically active stationary phases, *Agric. Biol. Chem.,* 44, 2987, 1980.
29. **Horiba, M., Kitahara, H., Takahashi, K., Yamamoto, S., Murano, A., and Ôi, N.,** Gas chromatographic determination of optical isomers of 2-(4-chlorophenyl)-isovaleric acid, *Agric. Biol. Chem.,* 43, 2311, 1979.

Chapter 4

MICROANALYSIS OF AMINO ACIDS BY GC/ECD AND GC/MS: APPLICATIONS IN ECOLOGICAL AND MEDICAL RESEARCH

Göran Odham and Göran Bengtsson

TABLE OF CONTENTS

I. INTRODUCTION

Amino acids constitute a readily accessible source of nitrogen and carbon for microorganisms, phytoplankton, invertebrates, and higher plants in a wide variety of environments. The nutrient nature of the amino acids is therefore of great importance in the multitude of interactions at play between different types of organisms. Furthermore, nutritional importance implies rapid metabolism and transformation.

The interplays between microorganisms and other cells constitute examples where the environments are extremely small and where the amino acids are consequently present in very low concentrations. Determination of amino acids, being either of the common type or the specific nonprotein type, therefore requires access to analytical methods of ultimate sensitivity and selectivity. Furthermore, the methods should preferably also enable measurements of amino acid transformation, e.g., by tracing the fate of specifically labeled molecules.

In this chapter emphasis will be placed on methods for handling low levels of material of biological origin, and high-resolution separation of amino acid derivatives using capillary columns and electron capture (EC) and mass spectrometric (MS) determination. In addition to being selective and sensitive, the MS technique is a prerequisite for possibilities of tracing stable isotopes, e.g., [15]N.

It is assumed that the reader has a basic knowledge of MS since the topics discussed essentially concern the quantitative aspects of the MS technique. Comprehensive information on general MS in biochemical and biological research is found, for example, in the collective volumes edited by Waller,[1] Waller and Dermer,[2] and Odham et al.[3]

II. MICROSCALE SAMPLE PREPARATION

A. Pretreatments and Cleanup

Two kinds of situations with respect to samples are frequently encountered in ecological and medical analytical work. In the first situation, the sample has a low concentration of analytes but the sample size is unlimited. A preconcentration step would facilitate determination of the compounds but also of trace impurities that will interfere with the compounds of interest. In the second situation, the sample size is limited but the concentration is sufficient to permit routine analyses of compounds, and the accuracy of the results is largely dependent on how representative the sampling technique is.

Our own work has often comprised a third situation, where the sample size is limited to some microliters or micrograms and the concentration falls in the parts per billion and parts per trillion range. This combination of limitations creates special demands for an accurate sampling technique, uncontaminated glassware and reagents, and sensitive and selective analytical equipment. Under this heading we will describe how such a microsample of amino acids can be prepared for analysis by GC.

Impurities on glassware and in reagents are a scourge in all kinds of microanalytical work. Innumerable hours have been spent in the lab tracing and eliminating a seemingly infinite number of sources of amino acid contamination. From our experience we can support two simple rules of thumb which will alleviate much frustration in the micropreparation of samples for GC/MS analysis:

1. Never mix your glassware with the common laboratory supply. Your glassware must be ultrapure and you can only trust yourself to keep it clean.
2. Do not trust the purity of a general supply of distilled and deionized water. Microorganisms grow well in distilled water pipings and in deionizers, and their exudates or excretions will ruin your reagent blank. Always use doubly distilled water from your own all-glass equipment and avoid microbial contamination of your containers.

Our procedure for sample treatment prior to GC analysis includes hydrolysis (for proteins and peptides), extraction (for the free amino acid pool of tissues, organisms, etc.), ion exchange, and derivatization. The whole procedure is carried out in Pyrex® glass capillary tubes, 80×2 mm I.D., and, alternatively, hydrolysis/extraction and ion exchange is performed with Teflon®-lined screwcap culture tubes, 14×90 mm. The tubes have to be soaked in a strong detergent, washed in double-distilled deionized water, and preferably also fired in an oven at 500°C to remove any trace of contaminants.

We have used the conventional procedure for protein hydrolysis, i.e., incubation of the sample at 110°C for 24 hr in 6 M HCl in an oxygen-free environment. With the capillary tube we use only 50 to 100 $\mu\ell$ of acid. The acid is added in excess with a microcap tube and the capillary tube sealed for incubation. To extract free amino acids, 70 to 80% ethanol is added to the tube and the sample is left in an ultrasonic bath for 2 hr. Tubes from hydrolysis and extraction are centrifuged for 15 min at 10,000 g and the supernatant is transferred either to another tube or to an ion exchange column. Some samples, e.g., those having a high content of organic carbon, may require a second cleanup step in addition to ion exchange to remove lipids and other neutral dissolved organic compounds. We have shaken the samples with ethanol-free chloroform without any losses of amino acids. Others[4] have used a combination of ligand exchange and cation exchange and encountered losses of glycine, glutamic acid, and aspartic acid.

The efficiency of two kinds of cation-exchange resins has been compared with respect to isolation and recovery of amino acids.[5] Dowex® 50W(H$^+$), 100/120 mesh, was efficient (99.77% of added amount of 20 amino acids recovered), whereas Amberlite® IR-120(H$^+$), 28/35 mesh, was unreliable and often gave a small residue of a viscous liquid in the capillary tubes after evaporation. The reason for this may be the influence of the surface area of the resin particles on the reaction rate constant of the ion exchange. The Amberlite® resin has a small surface area and the Dowex® resin a larger one. A similar success with a Dowex® 100/120 mesh cation-exchange resin was reported by others,[6-7] while Krutz[8] experienced difficulties in recovering amino acids using Amberlite® IR-120 (20/50 mesh) and Amberlite® IRA-400 (20/50 mesh). These difficulties probably also arose from the use of an anion exchange resin, which has not proven to be reliable for quantitative work with amino acids. Considerable losses of amino acids were also found with a combination of Dowex® 50 $\times$ 8 (200/400 mesh) and Dowex® 1 $\times$ 8 (200/400 mesh).[9] The same authors[9] used nitrogen-saturated resins to improve the recovery of the sulfur-containing amino acids, which may be partly oxidized. We have, so far, been unable to reproducibly determine minute amounts of cystine even with that precaution, and consequently reduced it to cysteine with SnCl$_2$ prior to ion exchange, as also suggested by Adams.[6]

The cation-exchange resin is washed in 3 M NH$_4$OH and 3 M HCl and kept in distilled water in a nitrogen atmosphere in a sealed flask. The ion exchange column, which is convenient to use for, e.g., enriched hydrolyzed samples, is prepared by adding 100 to 200 mg of resin to a glass funnel. The pH, which can be adjusted by acetic acid, is one of the main parameters in the displacement reaction on the sulfonic acid bed. The amino groups are partly protonated in acid solution, whereas the ionization of the carboxyl group is low. If an amino acid with a net charge of zero is passed through the resin, it will be adsorbed to some extent because of the acidic nature of the resin surface. Strong acidification is necessary to bring the monoamino acids completely into the univalent cation form. For example, at pH 2.5, 35% of phenylalanine, 66% of threonine, and 100% of the diamino acids are in the cationic form. In addition to cationic displacement reactions, amino acids are adsorbed to the resin by molecular adsorption, which is strong for amino acids, particularly those possessing an aromatic moiety.

Amino acids are adsorbed to the resin at a slow flow rate, 0.2 mℓ/min, whereas anions, sugars, lipids, and some polypeptides and proteins elute with distilled water. Amino acids are eluted with 3 *M* NH₄OH and the eluate is collected in a culture tube and dried in a freeze-drier or a rotary evaporator.

Instead of using a column, the resin may be added to the culture tube or the capillary tube (10 μℓ resin), then shaken with the acidified sample, and the excess liquid is discarded. The resin is then shaken with deionized distilled water in the tube, and the water is then discarded. Next, 3 *M* NH₄OH is added, the tube is shaken, and the "eluate" is withdrawn to another tube. In our experience this is a convenient and reliable procedure for microsamples with a small amount of amino acids, probably because the resin and reagent volumes can be kept to a minimum.

B. Derivatization

Several procedures to make amino acids volatile enough for GC analysis have been described and recently reviewed.[10] The most commonly used procedures require two steps of synthesis, i.e., esterification of the carboxyl group and acylation of the amino and other functions. The esters of butanol and propanol are particularly suitable with respect to volatility and chemical stability. We have achieved higher yield and better reproducibility using the *iso*-butanol derivatives than when using the *n*-propanol or *n*-butanol derivatives. This is probably due to a faster nucleophilic displacement by the branched reagent.

The acid catalyst may be introduced by allowing the alcohol to react with a small amount of acetyl chloride. We have found this procedure to be a convenient and accurate means of preparing the esterifying reagent and to be an excellent alternative to acidification of the alcohol with gaseous HCl.

Owing to its potential sensitivity, electron capture detection (ECD) was the choice for determination of microamounts of amino acids. Hence, one of the major demands on the derivative was its suitability for electron capture detection. Heptafluorobutyric anhydride (HFBA) readily acylates the amino groups, forming derivatives that possess high electron affinity, have reasonable thermal stability and volatility to allow rapid analysis, and, in addition, are also efficient for very small amounts of amino acids. One of the drawbacks of the heptafluorobutyryl (HFB) derivatives is the sensitivity to moisture. Hence, residues of HFBA, which rapidly deteriorate a GC column, cannot be removed by hydrolysis but have to be evaporated under vacuum, a procedure which has to be carefully standardized to prevent degradation of certain amino acids, particularly arginine. HFBA also has oxidative effects on the derivatives, especially on methionine and histidine, and an antioxidant, such as 2,5-di-*tert*-butylhydroxytoluene (BHT)[11] may be added to improve the reproducibility.

In our procedure, 50 μℓ *iso*-butanol/3 *M* HCl is added to the capillary tube, which is sonicated for 5 min and then incubated at 120°C for 20 min. Excess reagent is evaporated in a dessiccator at reduced pressure, then 25 μℓ dichloromethane is added to remove traces of moisture when evaporated. Acylation is performed under nitrogen in 25 μℓ of BHT (10⁻⁶ *M*) and 25 μℓ HFBA at 150°C for 10 min. Excess reagent is evaporated at reduced pressure and the residue dissolved in 5 μℓ of ethylacetate and stored in a sealed capillary tube at −20°C. We have tried to reduce the volume below 5 μℓ in the capillary tubes, but the reproducibility has not been satisfactory.

The principles of the procedures described here for cleanup and derivatization are used in many laboratories (see, e.g., Volume I, Chapters 2 to 6). Gehrke et al. thoroughly describe the meticulous care and special handling required for ultramicro analysis in their search for amino acids in returned lunar samples (Chapter 7, Volume II). They also present a discussion on sources of contamination and the appropriate laboratory techniques to eliminate amino acid contamination. Our main purpose has been to show that these procedures can be modified to allow microscale analyses. Any experienced technician can be trained to perform the sample preparations and keep contamination at a minimum.

III. GC SEPARATION

A. Sampling Techniques

In high-sensitivity analyses of amino acids the use of capillary columns is highly desirable. Not only do such columns provide superior separation of the individual components but, if properly optimized, the band broadening is much less than with packed columns. This means that the detection limit using any detection system may decrease by about one order of magnitude from packed to capillary columns (increased mass/time unit). To achieve this the injection technique should, however, be given careful consideration.

For obvious reasons the conventional and most commonly used technique involving splitting of the sample, whereby at least 90% of the sample is lost, is unsuitable. To overcome this general problem in high-sensitivity capillary GC, much effort has been made to find alternative techniques of sample introduction allowing analyses of the total sample. Two such techniques — splitless injection with or without "solvent effect" and "on-column" injection — both applicable to narrow-bore high-resolution columns, have received considerable attention and are extensively used. The former technique, particularly studied by Grob et al.,[12,13] Schomburg et al.,[14] and Jennings et al.,[15] involves a venting system in the injection port controlled by a valve. Following a conventional injection at low column temperature using a microsyringe, a split is put into operation by opening the valve after a certain period of time (usually on the order of 1 min). Hereby the solvent tail is vented out while the sample components to be analyzed proceed through the column. It is assumed that practically all of the analyte enters the column and that the actual chromatography begins when temperature programming of the column is started. By choosing the column temperature during the splitless injection and the boiling point of the solvent, the latter may act as a carrier of the analyte onto the column in such a way that a very narrow plug of material is formed ("solvent effect").

When analyzing complex amino acid derivative mixtures with a wide boiling point range, the linearity achieved by the injection system should be rigorously checked. The linearity is affected, as shown by Schomburg et al.,[16,17] for a number of experimental parameters including the volume and temperature of the injection port, the boiling point of the solvent, and the speed of injection. According to Schomburg et al.,[17] discrimination in the splitless injection of a C_{18}–C_{30} hydrocarbon mixture is increased by (1) too high a vaporization temperature, (2) contact of the needle with the injection port walls or fast heat transfer from the walls to the needle (hydrogen as carrier gas), (3) use of too volatile a solvent, and (4) too slow sample introduction with the syringe.

On-column (direct) sampling is generally considered to be superior to splitless sampling, especially for analyses where discrimination constitutes a serious problem. An injection in which the sample is placed directly within the unheated column should in theory eliminate the possibility of discriminatory injections. The principal experimental difficulty in on-column injections is the design of devices where syringes or pipettes that fit inside even a 0.20-mm I.D. column can be used. A number of on-column injection systems utilizing septum-free valved inlet systems have been described (e.g., Reference 17). The matter of choice depends largely on the sample volumes to be introduced.

Moodie and Burger[18] compared on-column injection with split injection of a standard mixture of 18 amino acids on the order of 10 ng each and found that precision of results was significantly better using on-column injection with a serious discrimination of larger molecules when split injection was used. Coefficients of variation (CV) exceeded 4% for the last seven eluted peaks. Histidine was not detected unless dissolved in acetic anhydride with an anything but encouraging CV at 27%. Hence, split injection is quantitatively unreliable at least for the higher boiling amino acids, whereas on-column injection is a good choice except for the higher probability of a rapid column degradation due to solvent

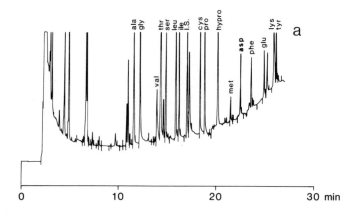

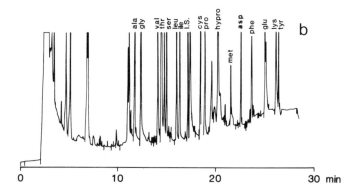

FIGURE 1.   GC of a splitless injection of a mixture of amino acids (approx. 500 pg each) on a 25 m (I.D. 0.25 mm) fused silica column coated with SE-54 and connected to an ECD. (a) 15 s splitless period. (b) 30 s splitless period, which was found to be sufficient for quantitative determination. Injection temperature, 220°C; detector temperature, 300°C; temperature programming 95°C for 5 min, 6°/min to 210°C. Nitrogen was used as carrier gas at 1.8 mℓ/min and split ratio was 1:20.[50]

condensation. The same is true for splitless injection with solvent effect, when the oven temperature is lowered to 30 to 40°C for HFBA derivatives dissolved in ethyl acetate. With proper selection of the splitless period and split ratio at venting, amino acids can be quantitatively determined using splitless injection on a narrow-bore fused silica column (Figure 1), although splitless injection is more precise with support-coated open tubular (SCOT) columns with a higher sample capacity (see Section III.B).

B. Capillary Columns

There are two types of capillary columns to be considered in high-resolution GC of amino acids. These include the SCOT columns and the wall-coated open tubular (WCOT) columns. The SCOT columns are made from glass, possess a comparatively large inner diameter, typically 0.5 mm, and are supported on the inner walls by vitreous 7 μm (average) silica particles which together with the wall itself form the foundation for the stationary phase. The SCOT columns have inferior separation qualities compared to the WCOT columns (about half the number of theoretical plates per unit length) but possess sample capacities that allow direct syringe microinjections in the manner used for packed columns. Since there is no need for elaborate and time-consuming GC manipulations, as is the case in the splitless

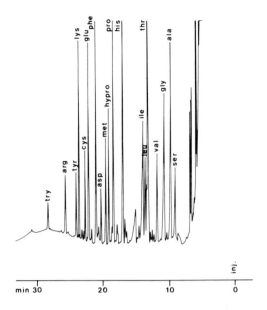

FIGURE 2. GC of a reference mixture of 19 protein amino acid *N*-HFB-*i*-butyl derivatives on a 50-m SCOT column with ECD using direct injection. Each peak represents 200 to 600 pg of amino acids.[27]

or on-column injection modes (see Section III.C), the SCOT columns are particularly useful for routine analyses or for larger series of samples. The diminished separation qualities may be compensated for by increasing the length of the column. A typical gas chromatogram with ECD involving direct injection on a wide-bore 50-m SCOT glass column (OV-101 and OV-17 as stationary phases) is reproduced in Figure 2.

For best separation, narrow-bore (I.D. 0.2 to 0.3 mm) WCOT columns are the choice. These may be either of the glass or fused silica type. Glass columns possess the advantage of allowing modifications of the inner surface, after which practically any stationary phase can be supported, whereas fused silica columns at present are suitable only for a restricted number of non- and medium-polar phases, with merits of great flexibility and easy handling on a routine basis. The authors' experience of GC of amino acids is restricted to the use of the relatively nonpolar fluoroacyl alkyl ester derivatives which separate well on the common polysiloxane stationary phases. We prefer SE-54 (1% vinyl, 5% phenyl methyl) coated on persilylated fused silica columns. Columns with immobilized ("bonded") stationary phases may be advantageous for the separation of HFB-*i*-butyl derivatives of amino acids. The long-term deleterious qualities of the derivatization reagents and their degradation products on the column can be overcome by occasional back flushing with a suitable volume of solvent.

C. Capillary Columns with Chiral Stationary Phases

As a result of the recently developed chiral stationary phases for capillary GC, possibilities to examine interactions involving prokaryotic organisms (bacteria) are open since these organisms contain unique amino acid enantiomers such as D-alanine and D-glutamic acid in defined proportions in their cell walls. Separation and quantification of amino acid enantiomers are discussed in detail in Volumes II and III.

A variety of different chiral stationary phases, e.g., α-hydroxycarboxylic acid esters,[19] *N*-trifluoroacetyl (TFA)-L-phenylalanyl-L-leucine cyclohexyl ester,[20] and *N*-docosanoyl-L-valine *tert*-butyl amide,[21] have been successfully used. More recently, high-temperature-

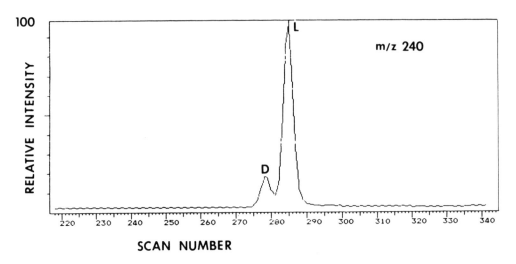

FIGURE 3.   Mass fragmentogram of the *N*-HFB-*i*-butyl ester derivative of D- and L-alanine monitoring at *m/z* 240 (EI, 70 eV, Ribermag R-10-10c GC/MS/DA system).[25]

stable stationary phases have been prepared by grafting chiral amide phases onto modified polysiloxane backbones.[22-24] The latter approach, apart from leading to particularly temperature-stable phases, is attractive in that it enables one to coat the chiral phases on fused silica capillaries. Figure 3, showing a MS determination of the D:L ratio of alanine in intact cells of Group A streptococci type M 15 using a 25-m glass capillary column coated with L-valine-*tert*-butyl amide coupled with a polysiloxane as stationary phase, may serve as an example.[25]

IV. DETECTION AND QUANTIFICATION

A. ECD

One of the most attractive volatile amino acid derivatives is the *N*-HFB alkyl ester. It forms relatively stable derivatives which are suitably volatile for GC analyses and which, in addition, give very high responses with ECD. The ECD is a highly specific detector operating on the principle that certain types of molecules are electron absorbers and readily form negative ions when exposed to a concentration of free electrons. Free electrons are produced by ionization of the carrier gas molecules by energetic β-particles emitted from a radioactive foil in the detector cell. Electron-absorbing sample molecules introduced into the detector cell by the effluent of the column cause a decrease in the free electron concentration — a decrease which is sensed by the electronic circuitry. Modern ECDs are operated in a pulse-modulated way whereby the ECD becomes an integral component of an electronic feedback network. Pulsed ECDs offer in general a wide linear range ($>10^4$), a quality which is of particular importance in the analyses of complex amino acid mixtures having a wide range of individual component concentrations.

The high sensitivity of the HFB amino acid derivatives toward the ECD is attributed to the high electron-withdrawing capacity of the acyl group which is assumed to depend on the increased polarity of the carbonyl carbon atom induced by the electronegativity of the seven fluorine atoms.[26]

The absolute sensitivity of the ECD for the individual amino acid derivative varies considerably however. Diacylation, e.g., in the case of serine, tyrosine, histidine, and tryptophan increases the sensitivity (Table 1).[27] Furthermore, the presence of the imidazole and indole ring systems in histidine and tryptophan increases EC resulting in high relative molar response

Table 1
EFFECT OF ECD ON AMINO
ACID RMRs

Amino acid	$RMR_{ECD}:RMR_{FID}$
Ser	2.0
His	2.5
HyPro	1.8
Met	1.4
Glu	3.9
Tyr	1.4
Try	6.0

Note: RMRs are calculated for each amino
acid relative to norleucine by peak-
height measurement. Sensitivity of ECD
and FID is compared by using the quotient between RMRs. Quotients > 1 are
given. RMRs for other amino acids tal-
lied with those given in Reference 28.

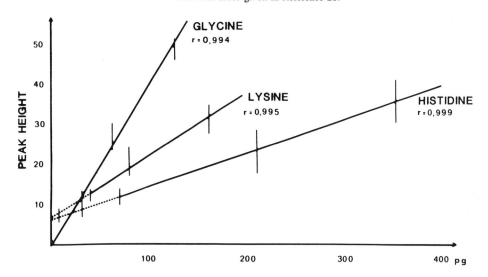

FIGURE 4. Peak heights (mean and SD) vs. sample size for the *N*-HFB-*i*-butyl derivatives of glycine, lysine, and histidine using ECD. (Varian® model 3700 GC employing a 50-m SCOT glass capillary column and direct injection.)[27]

(RMR) to norleucine quotients. An unexpected difference in sensitivity between aspartic (quotient < 1) and glutamic acids (quotient 3.9) is also noted. Possibly this depends on the ready formation of the cyclic pyrrolidone carboxylic acid ester from the diester of the latter at the esterification temperatures used.[29]

Bengtsson and Odham[27] found the linearity of the ECD satisfactory (r = 0.99) within the 10- to 400-pg range (Figure 4) and the detection limit for the HFB alkyl derivatives was less than 1 pg for most amino acids. This corresponds to a detection limit about three orders of magnitude less than that obtained by a flame ionization detector (FID).

There are two obvious disadvantages in ECD analysis of trace amounts of amino acids. The first one is the appearance of extraneous small peaks in a chromatogram. Some of these peaks represent impurities in the amino acid standard (very few of the commercial salts we have used were of 100% purity) (cf. Figure 2), others represent impurities in glassware, reagents, and sample, and some of them probably denote incomplete diacylation of the amino acids. Another source of extraneous peaks is the presence of carbonyl compounds, e.g., butyraldehyde, in *iso*-butanol which may condense with the NH_2 group during butylation.

The second disadvantage is the reduction in the standing current of the detector due to bleeding from the stationary phase during temperature programing. The reduction may arise from trace contaminant vapors in the carrier gas or pyrolysis products of the stationary phase. Bleeding increases exponentially with temperature, producing a dramatically rising baseline at higher oven temperatures, which can make quantitation of peaks difficult. The electrometer voltage can be modified to compensate for the baseline drift[30] (Figure 5).

B. Repetitive Scanning of Spectra and Extracted Ion Current Profiles

Most computerized modern mass spectrometers, either of the magnetic or quadrupole types, have the capability for repetitive scanning of mass spectra. The total signals given by the instrument for a chosen scan range during each scan are stored in the external memory of the computer and are usually, while the analysis is performed, visualized in real time on a screen. The plot thus represents, as a function of time, changes in total ion currents. For a quadrupole mass spectrometer, typical scanning times are 1 msec per mass unit, which often means that a complete scan is recorded during less than 1 sec. The introduction of data acquisition systems havₗᵥ g large storage capacities has made feasible the recording of complete chromatograms with the GC as inlet system by repetitive scanning. The technique originally reported by Hites and Bieman[31] is often described as mass chromatography.

Figure 6A illustrates a mass chromatogram of the *i*-butyl-HFB-derivatives of the amino acids obtained by acid hydrolysis of 3.2 μg of freeze-dried *Myxococcus virescens* using a 25-m fused silica GC column coated with SE-54. Ions of selected *m/z* characteristic for certain amino acids can be extracted, in real time, from the stored scan data. The result of this facility, often known as "extracted ion current profile" (EICP), is particularly useful in the screening of unknown complex amino acid mixtures. Figures 6B to D show EICP of ions characteristic of the HFB *i*-butyl derivatives of methionine (m/z 131 = $H_3CSCH_2CHCHCO_2$ or $OCSCH_2CHCO_2$), leucine and *iso*-leucine (m/z 282 = M-$OCOC_4H_9$), and tyrosine (m/z 360 = M-($OCOC_3F_7$ + OC_4H_8)), respectively.[32] It should be emphasized that EICP is merely an assortment of ion intensities by the computer, so that the achieved sensitivity is no greater than that obtained by the repetitive scanning procedure. Nevertheless, sensitivity is comparable with that of the conventional FID because with modern mass spectrometers reproducible electron impact spectra, particularly when using capillary GC as inlet system, can be obtained from as little as 1 ng of an individual amino acid.

C. Selected Ion Monitoring

Although EICP is valuable for screening of mixtures of unknown compositions, the sensitivity achieved is only in the nanogram range. However, by allowing the mass spectrometer to focus only on preselected ions (SIM), thus maximizing the duty cycle of the instrument, the sensitivity may be increased by several orders of magnitude. The degree of sensitivity will depend on the relative intensity of the ion chosen for monitoring and the type of ionization used, e.g., electron impact ionization (EI) or chemical ionization (CI). The ultimate response is obtained when the measurement is performed continuously at only one ion species (single ion monitoring). Like EICP, SIM offers excellent specificity, but the main drawback is that a decision on what compounds to be determined has to be made prior to the analysis.

Computerized switching of masses may be performed with both magnetic and quadrupole instruments. Technically, it is easier with the latter type and several such instrumentations allow selection of any *m/z* within the mass range.

Figure 7 illustrates the selective determination by single ion monitoring of valine produced

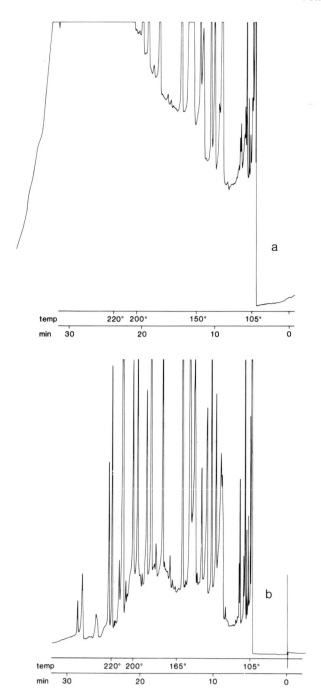

FIGURE 5.   GC after injection of a 0.5-μℓ sample of *N*-HFB-*i*-butyl derivatives of reference amino acids on a 50-m SCOT column with temperature programming and ECD; without (a) and with (b) baseline correction (Varian® model 3700 GC).[30]

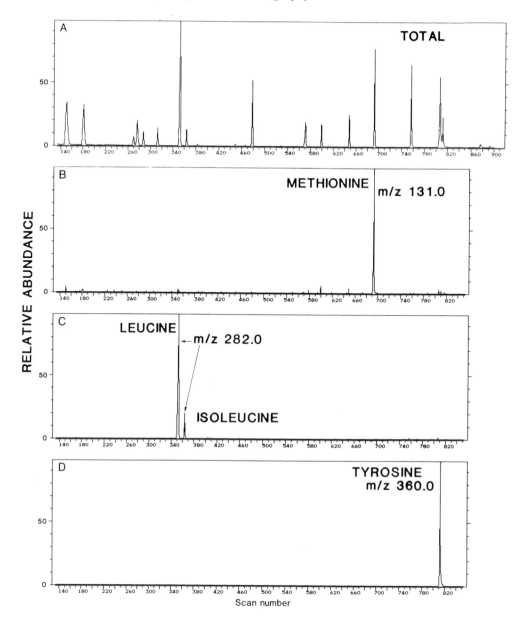

FIGURE 6. Mass chromatograms of the *N*-HFB-*i*-butyl derivatives of the amino acids obtained by acid hydrolysis of *Myxococcus virescens* cells. (A) Total ion current; (B to D) EICP of the ions *m/z* 131 (methionine), *m/z* 282 (leucine and isoleucine), and *m/z* 360 (tyrosine). GC and MS conditions (cf. Reference 33).[50]

by hyphomycetes on leaves in a running water system. The detection limit when monitoring in EI at *m/z* 268 (= M-OCOC$_4$H$_9$) was determined by the general background level at the laboratory, 0.8 pg.[5]

D. MS Quantification

The general principle of calibration of the mass spectrometer for quantitative determination is that known amounts of the analyte are introduced and the resulting MS signal measured. In computerized MS systems calibration plots based on linear regression techniques are readily obtainable. A typical such plot for the HFB *i*-butyl-derivatives of glycine, lysine,

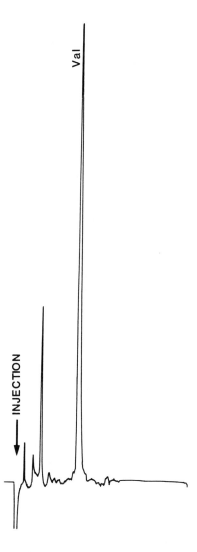

FIGURE 7. Identification of valine (*N*-HFB-*i*-butyl derivative in a water sample from a Swedish creek (Stampenbäcken) using single ion monitoring at *m/z* 268. The signal represents about 50 pg of valine (EI, 70 eV, Varian® model 112 GC/MS system)[5].

and histidine using SIM is shown in Figure 8.[27] The linearity of the plot indicates that direct comparison of MS signals can be used for quantification. The illustrated plot was, however, based on authentic amino acids and does not consequently take into account losses occurring during, e.g., extraction, hydrolysis, and cleanup of natural samples.

The calibration plots for lysine and histidine in Figure 8 indicate a finite Y-intercept. The result reflects the difficulties of removing the normal background amounts of these common amino acids during the derivatization procedure.

To compensate for losses during sample preparation and variabilities in the MS system, such as fluctuations of the pressure in the ion source and changes of the electron multiplier gain, an internal standard should be added to the analyte as early as possible in the analytical process. Prior to the measurements the usual calibration curve is constructed by using the standard/reference weight ratio and response ratio.

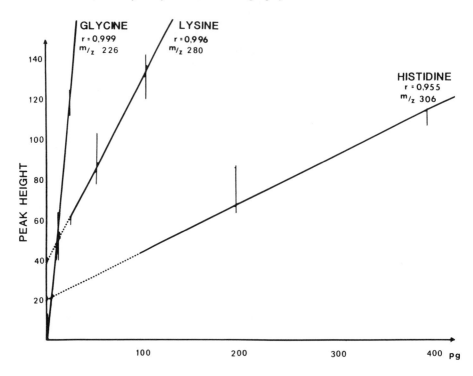

FIGURE 8.   Peak heights (mean and SD) vs. sample size of HFB-*i*-butyl amino acids in MS. The derivatives were separated on a 50 m SE-30/OV-17 SCOT-column and detected by SIM on a Varian® MAT-112 mass spectrometer.[27]

Since the MS signal depends both on the chemical structure and the amount of the analyte, the choice of internal standard is important. A structural analog to the analyte is commonly used in conventional GC. Such an analog can also be used in GC/MS work provided it contains the same structural elements that cause fragmentation after ionization. Norleucine and pipecolic acid have been used comprehensively by several workers for quantification of the protein amino acids. 8-Aminooctanoic acid was preferred, primarily for reasons of a suitable GC retention time, by Tunlid and Odham[33] in studies focusing on determinations of the bacteria cell-wall-specific muramic and diaminopimelic acids. Almost ideal internal standards are the optical antipodes, which are true chemical duplicates of the L-amino acids, provided both isomers can be separated on a chiral stationary phase. This was demonstrated by Frank et al.[34] for determination of amino acids in blood serum, and is described in Chapter 2 of Volume II.

The MS detector is unique in its ability to measure masses. This allows use of the optimal stable-isotope-labeled analog which from both chemical and MS viewpoints behaves very similarly to the corresponding analyte. For amino acids ^{2}H(d), ^{13}C, and ^{15}N are possible candidates for labeling. Deuterium, being the most inexpensive isotope, is the most used marker,[35] but along with the recent reduction in cost of ^{13}C-enriched material this labeling has increased interest. The use of ^{15}N in quantitative MS is discussed in Section IV.E.

The increase in masses in the standard due to labeling deserves consideration. Because of the distribution of isotopes in nature, the heavy isotopes ^{13}C, ^{2}H, ^{15}N, and ^{17}O contribute to the mass spectrum in such a way that ions of the nominal mass *m* are measurably accompanied by isotopic ions of at least *(m + 1)* and *(m + 2)*. Therefore, to reduce the blank level, and hence the detection limit, a mass increase of three or more in the standard should be attempted.

Quantification of trace amounts of glutamic acid in water, involving the use of 2,3,3,4,4-

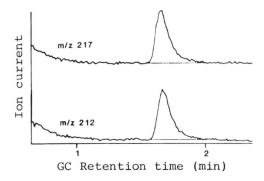

FIGURE 9.   Mass fragmentogram of *m/z* 212 (glutamic acid) and *m/z* 217 (d$_5$-glutamic acid) of a water sample.[36]

d$_5$ glutamic acid as internal standard, is shown in Figure 9.[36] Using SIM with EI, the fragments of *m/z* 212 and *m/z* 217 (standard), formed by loss of a methoxycarbonyl group from the dimethyl ester *N*-trifluoroacetylated acid, are recorded. Notably, the difference in GC retention time between the analyte and internal standard is negligible.

The availability of suitably labeled amino acids for use as internal standards is frequently a limiting factor. The problem can be overcome by incorporating the isotope during the derivatization step, e.g., by using 4,4,4-d$_3$ butanol as esterifying reagent in the preparation of the standard. However, possible losses during manipulations prior to the derivatization procedure are hereby not under control.

High sensitivity GC/MS of amino acids of biological origin often involves a number of intricate manipulations such as hydrolysis, purification by ion exchange, and multistep derivatizations. Therefore, it is important to take the overall manipulation error into account and, if necessary, make corrections for it. The error of the analytical procedure can be tested by adding a known amount of a structural or labeled analog of the analyte to the original matrix and determining the recovery. The yield in this highly recommended test serves as an indication of the losses during the analytical procedure.

The precision in quantitative MS, expressing the reproducibility of a measurement, is unfortunately a grudgingly treated subject and when data are given in the literature, they usually relate to only one MS apparatus. A model for the precision of an analytical MS method can be exemplified by data given in Table 2 taken from a study of ^{15}N incorporation into dog plasma amino acids during continuous infusion of ^{15}N-leucine.[37] The isotopic enrichment in 15 protein amino acids was determined using GC/MS adapted for SIM with CI (methane). The precision defined as s.d./M $\times$ 100 (s.d. = standard deviation, M = mean of the measurements) was below 2% for each isotope ratio measurement. On the whole, this value appears representative for current MS methodology. However, at the detection limit of the instrument, precision often decreases significantly.

Many factors influence the sensitivity that can be achieved in quantitative MS of amino acids. These include the chemical nature of the analyte and the chromatographic and MS conditions. Specifications of sensitivity and detection limit are therefore meaningful only if the analytical and instrumental setups are thoroughly defined.

Since the MS response is defined by the number of ions per unit time, it is important in highly sensitive work to aim at sharp GC effluents. Capillary columns are outstanding in this achievement and should consequently always be chosen in such investigations. Needless to say, splitless or on-column techniques are imperative. Furthermore, low column bleed is also of importance.

With respect to the MS conditions, the type of ionization and the ionization energy are of decisive importance. The sensitivity that can be attained for a given amino acid using

Table 2
¹⁵N INCORPORATION INTO DOG PLASMA
AMINO ACIDS DURING CONTINUOUS
INFUSION OF (¹⁵N) LEUCINE

Amino acid	m/z monitored[a]	Isotope ratio at 0 hr, $\times$ 100[b]	Ratio difference 9 hr to 0 hr, $\times$ 100[c]
Ala	(201, 202)	11.236 ± 0.047	11.872 ± 0.062
Val	201, 202	11.354 ± 0.064	12.054 ± 0.107
Gly	(188, 189)	10.242 ± 0.058	10.344 ± 0.066
Ile	216, 217	12.462 ± 0.063	13.464 ± 0.106
Leu	216, 217	12.450 ± 0.091	18.110 c
Pro	200, 201	11.402 ± 0.095	11.506 ± 0.050
Thr	246, 247	12.460 ± 0.133	12.612 ± 0.053
Ser	232, 233	11.370 ± 0.055	11.636 ± 0.062
Asp	260, 261	13.658 ± 0.093	13.748 ± 0.074
Met	234, 235	12.338 ± 0.072	12.350 ± 0.037
Phe	250, 251	15.672 ± 0.044	15.760 ± 0.060
Glu	(302, 303)	16.684 ± 0.123	17.030 ± 0.068
Try	308, 309	18.072 ± 0.270	18.072 ± 0.115
Orn	259, 260	13.914 ± 0.096	14.234 ± 0.066
Lys	(301, 302)	17.224 ± 0.098	17.144 ± 0.134

[a] m/z values in parentheses are the $(M + C_2H_5)^+$ ion pair rather than the $(M + H)^+$ ion pair.
[b]   Mean ± standard deviation; n = 5.
[c]   For n = 2 rather than n = 5.

SIM is dependent on the total ion current represented by the monitored ion. Fortunately, the commonly used amino acid derivatives usually produce characteristic ions that are proportionately abundant in the high mass range when EI is used.[38,39] For most amino acids the detection limit using single ion monitoring in EI at 70 eV falls in the very low picogram range. In certain cases it is even possible to reach the femtogram range.[5]

The MS fragmentation pattern is highly influenced by the use of CI, resulting in very little fragmentation. The adduct ions formed, the nature of which depends on the reactant gas used, are usually related to the entire amino acid molecule and furthermore represent the majority of the total ions. CI is, however, a much "softer" method than ionization with 70-eV electrons. The total ion current is therefore considerably less with CI than with EI and consequently, whether or not sensitivity is increased by using CI, must be experimentally determined for each individual amino acid.

It was mentioned that SIM is often used for selected measurements in the picogram range. Such quantities are too small to allow registration of complete mass spectra to ensure the identity of the analyte. In general, specificity in SIM decreases with decreasing mass number of the monitored fragment. Higher mass of the ions being recorded reduces possible interference from instrumental background and from column bleed. Due to the nature of CI, it may therefore be stated that, in general, specificity is better in CI than in EI.

E. Measurement of ¹⁵N:¹⁴N Ratios in Amino Acids

There is an increasing demand for detailed studies of the flux of nitrogen in many biological and biochemical systems. One approach to such investigation comprises the use of ¹⁵N-labeled precursors. The traditional way, based on the pioneering work of Rittenberg,[40] involves MS determinations of the ¹⁵N abundance of amino acids by isolation of the individual acid followed by quantitative oxidation of the amino or amido groups to N_2.

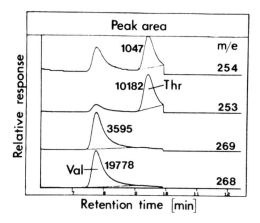

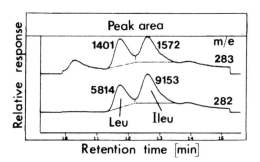

FIGURE 10. SIM of the *N*-HFB-*i*-butyl derivatives of a complex standard amino acid mixture containing 125 pmol of each compound. Selected ion monitoring parameters were *m/z* 268:269 (valine), 282:283 (leucine and isoleucine), and 253:254 (threonine). The integrated ion signals are shown. All amino acids were assumed to be at natural ^{15}N abundance (0.37% atom% ^{15}N).[42]

Introduction of N_2 into the ion source and measurement of the ratios of the ions of *m/z* 28, 29, and 30 and careful corrections for the nitrogen present in the residual air, will in principle yield the ^{15}N abundance of the amino acid. The methods are, however, time-consuming when applied to a large number of amino acids and furthermore, in practice, samples of less than 10 μg of N cannot be subjected to ^{15}N analysis with any degree of accuracy.[41] Recent investigations involving both EI and CI have therefore evaluated the possibilities of determining the ^{15}N:^{14}N ratios in amino acids by direct studies of suitable fragments in the MS spectra of derivatized amino acids using GC/MS/DA instrumentation.

The careful investigation by Rhodes et al.[42] comprising the kinetics of $^{15}NH_4^+$ assimilation in *Lemna minor* by GC/MS using EI of the HFB-*i*-butyl esters of 11 individual amino acids illustrates the situation. The authors showed that selected ion monitoring of major N-containing fragment ions of *m* and *(m + 1)*, can be used to calculate ^{15}N abundance with an accuracy of ± 1 atom% excess ^{15}N using samples containing as little as 30 pmol of individual amino acids. As an example, the SIM of a complex mixture of amino acid standard derivatives (packed column) showing two typical selected fragmentograms of alanine *(m/z* 240, 241), valine *(m/z* 268, 269), threonine *(m/z* 253, 254), and the leucines *(m/z* 282, 283) are reproduced (Figure 10).

Also, the possibility of using low-resolution MS operating in the SIM mode with CI has been evaluated.[43] Using a packed column with *iso*-butane as the reagent gas the authors

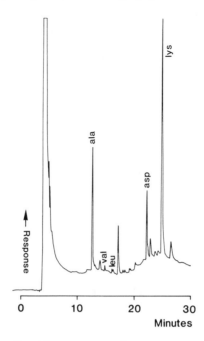

FIGURE 11.   GC of amino acids excreted
from the fungus *Arthrobotrys oligospora*. The
amino acids peaks represent 9 to 4000 ng on
a 25-m WCOT-column coated with OV-101.[5]

concluded that the GC/MS method could discriminate between mean levels separated by
0.1 atom% [15]N with 68% confidence at the 0.1-nmol level. Quantification was performed
by measuring the protonated molecular ions.

Using either method of ionization it may be anticipated that the use of capillary columns
will significantly increase the sensitivity in terms of individual samples.

V. MICROENVIRONMENTAL APPLICATIONS

A. Ecological Applications

Relatively few attempts to use microanalytical GC or GC/MS for amino acid determinations
in ecological or medical research have been made. This moderate use of a powerful analytical
technique probably has its roots in a traditional predominance for other methods of amino
acid analysis, e.g., automatic amino acid analysis using ion exchange resins, enzymatic
assays, paper chromatography, and high-voltage paper electrophoresis, and a general lack
of interest in problems that would require ultramicroanalysis. It is our hope that demonstration
of the usefulness of GC/ECD and GC/MS for the determination of trace amounts of amino
acids from biological microenvironments will encourage more laboratories to aim at equipment enabling such analyses.

One of the major research areas at our laboratory is microbial interactions, with special
emphasis on chemical communication and regulation of population growth and colonization.
Traditionally, microbial populations are kept in laboratory cultures at high densities, and
isolation and quantification of metabolites present no problem with conventional techniques.
For example, a population of the nematophagous fungus *Arthrobotrys oligospora* shows an
exudation of free amino acids in the order of 0.1 to 4 μg/hr (Figure 11)[5] when grown
axenically in the laboratory, but interpretation of the result may be questionable, since the
exudation may simply be density-dependent and a consequence of favorable growth conditions. It is known that nitrogen-containing exudates from nematodes play a role in the trap

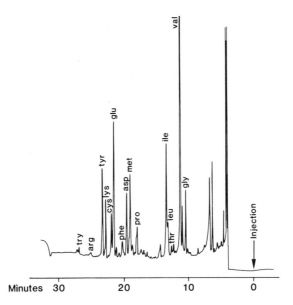

FIGURE 12. GC of amino acid exudation from one axenically grown specimen of the nematode *Panagrellus redivivus* incubated for 19 hr in deionized water. The peaks represent between 50 pg (threonine) and 2500 pg (valine) corresponding to an exudation rate of 1 and 50 ng · hr^{-1}, respectively.[27]

formation of this fungus,[44] and since it is likely that one single nematode can induce trap formation, we demonstrated that one individual *Panagrellus redivivus* exuded significant amounts of free amino acids into deionized water in a capillary tube (Figure 12).[27]

Aphids are known to pierce the host's phloem and extract nutrients which are then released as honeydew, mainly sugars, which subsequently falls to the ground. Although detrimental in several ways, aphids are also assumed to be beneficial for the plant by honeydew production. We found that one droplet of honeydew from an aphid on a black currant leaf contained large amounts of free amino acids (Figure 13), and it was calculated that one single leaf contributed 2 µg of amino acids per day to the soil.[27]

The exudation of free amino acids from the leaves of a small herbaceous plant, the lady's mantle, *Alchemilla vulgaris,*[27] can be estimated by collecting dew droplets from its surface. Amino acids in a freshly formed evening dew droplet (50µℓ) was compared with those in a morning dew droplet and we found that the concentration increased considerably during the night (Figure 14). The dew droplet thus represents a rich source of nitrogen readily accessible to phylloplane microorganisms.

Natural waters — whether marine or limnic — have at the air/water interface a very thin layer consisting of enriched organic and inorganic material in which populations of various microorganisms live. The free amino acid content of this surface microlayer was compared with the subsurface water from a lake (Figure 15), and an enrichment corresponding to several orders of magnitude in the microlayer was observed.[27]

Estimation of microbial biomass by measuring specific biochemical constituents of the microbes has been suggested as a sensitive and selective alternative to conventional plate counting and direct counting. Several characteristic components are found in the peptidoglucan moiety of the cell walls of bacteria. The amino sugar *N*-acetyl-muramic acid (MuAc) is a constituent of nearly all prokaryotic organisms. Diaminopimelic acid (DAP) is present in probably all Gram-negative and in some Gram-positive bacteria. Paper chromatography, thin-layer chromatography, ion exchange chromatography, enzymatic analysis, and GC and

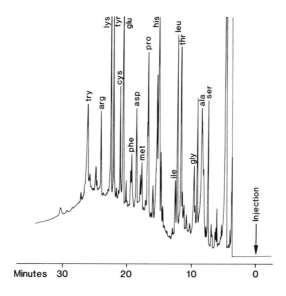

FIGURE 13.   GC of amino acids in honeydew from aphids on a black currant leaf. The amounts detected ranged from 130 pg (glutamic acid) to 3 ng (proline).[27]

GC/MS have previously been used to determine MuAc and DAP. The described GC micromethod was used in our laboratory to quantify MuAc and DAP in *Escherichia coli*.[33] Whole bacterial cells were hydrolyzed and peaks corresponding to MuAc and DAP appeared in the chromatograms (Figure 16A). The mass spectrometer was then used to monitor mass fragments of m/z 380 (DAP) and m/z 383 (MuAc) (Figure 16B). The detection limit was about 10 pg using both ECD and SIM which corresponds to approximately 30 ng (dry weight) of *Escherichia coli* in a 5-$\mu\ell$ sample.

B. Medical Applications

It is now recognized that sensitive and specific GC/MS methods can be used as diagnostic tools in medical microbiology.[45,46] Conventional isolation of microorganisms is sometimes tedious and unreliable and identification of species or type-specific metabolites by GC/MS is a fast and promising alternative. As an example, we have demonstrated contamination of tissue cell cultures by mycoplasma. The confirmation is based on the ability by at least some species of mycoplasma to catalyze enzymatic hydrolysis of arginine to citrulline and ornithine, a phenomenon which is not known for mammalian tissues. Ornithine exuded by *Mycoplasma hominis* was determined after 2-hr incubation by SIM detection monitoring at m/z 266 (M-($C_3F_7CONH_2 + OCOC_4H_9$)) (Figure 17).[27]

The combination of microdissection and mass fragmentography has proven useful for studies on neurotransmitters at the synaptic level. Bertilsson and Costa[47] developed a method for the simultaneous quantitation of glutamic acid and γ-aminobutyric acid (GABA), which is known as an inhibitory transmitter with glutamic acid as a precursor, by mass fragmentography. The two amino acids were derivatized in a one-step reaction with pentafluoropropionic anhydride and hexafluoroisopropanol using deuterium-labeled glutamic acid and GABA as internal standards. The presence of endogenous GABA was established in the rat cervical ganglion by mass fragmentography in small punches, about 50 μg of protein, of brain nuclei.

Wolfensberger and colleagues[48,49] in Zürich have developed a method to collect perfusate in the optic tectum of the pigeon brain and designed a micromethod to determine the presence

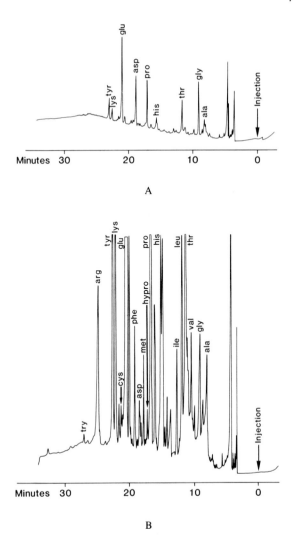

FIGURE 14.   GCs of amino acids in a dew droplets of lady's mantle from an evening (A) and a morning (B) sample.[27]

of glycine and glutamic acid in the perfusate. The amino acids were released from the optic tectum by electric stimulation and then determined by mass fragmentography of their *N*-pentafluoropropionyl hexafluoroisopropyl esters using deuterium-labeled amino acids as internal standards[48] (Figure 18) and by glass capillary GC in combination with thermoionic detection[49] (Figure 19). Their findings strongly support the hypothesis that glycine is an inhibitory neurotransmitter in the pigeon optic tectum.

VI. CONCLUSIONS

The size of an organism and its habitat limits the rate of accumulation of information on its ecology. For example, although electrodes for pH measurements have been known and used for decades, there is still substantial uncertainty about pH in the soil close to a bacterial cell and a fungal hyphae or spore because of the inherent difficulties to measure pH in a microhabitat. Difficulties arise not only because of the limited habitat size and its heterogenous nature, but also because of lack of microanalytical methods for the determination.

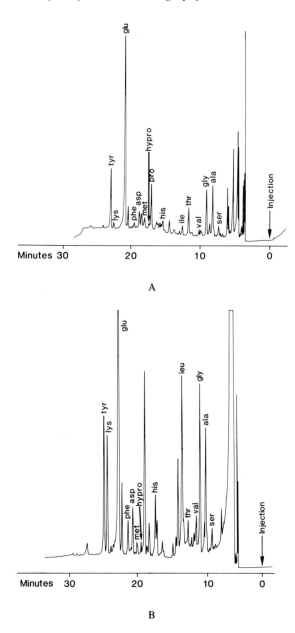

FIGURE 15.   GCs of amino acids in subsurface water (a) and in the surface film (b) from Lake Skäravatten, Sweden. The peaks in (a) represent 30 (lysine) to 400 pg (glycine) corresponding to 0.08 to 1.2 ng · mℓ^{-1} subsurface water. The peaks in (b) represent 50 (hydroxyproline) to 2000 pg (e.g., glycine) of amino acids corresponding to 0.3 to 9.5 ng/cm^2 surface film.[27]

The procedure described in this chapter for the ultramicrodetermination of amino acids by GC/ECD or GC/MS opens new possibilities to study microbial interactions in a diversity of microenvironments. Providing satisfactory sampling methods can be designed for a specific environment, concentrations in the ppb or ppt range of amino acids in a sample size of some few micrograms or microliters can be reproducibly handled in capillary tubes through a cation exchange cleanup step and a derivatization to the *N*-HFB-*iso*-butyl ester amino acids.

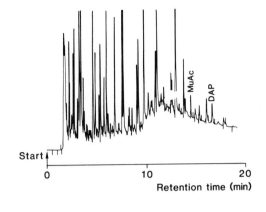

A

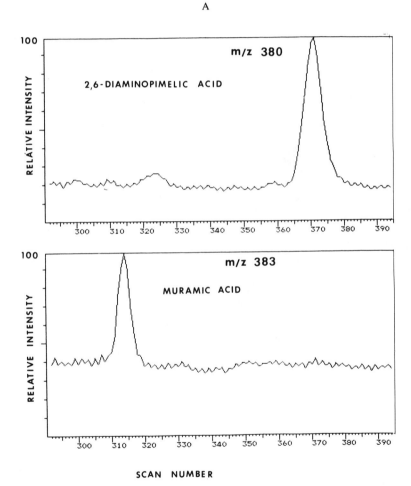

B

FIGURE 16. (a) GC of amino acids in a hydrolysate of an *E. coli B* culture. The peak corresponding to muramic acid (MuAc) represents about 130 pg and that to diaminopimelic acid (DAP) 140 pg. (b) Mass fragmentograms of muramic acid and diaminopimelic acid in the same hydrolysate. The signal corresponding to MuAc represents about 580 pg and that to DAP represents about 700 pg.[33]

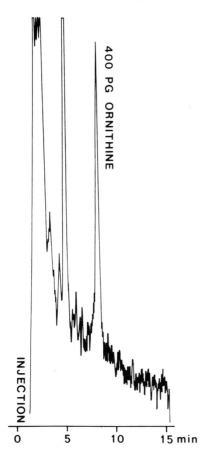

FIGURE 17.   Mass fragmentography of ornithine produced by *Mycoplasma hominis*
during incubation in a phosphate buffer.[27]

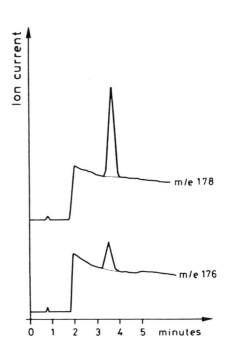

FiGURE 18.   Mass fragmentogram of the PFP-
HFIP derivatives of glycine (*m/z* 176) and d_2-
glycine *(m/z* 178) extracted from perfusates. The
signal for d_2-glycine corresponds to approximately 2.6 pmol.[48]

Modern GC and MS techniques, including splitless or on column injection, capillary col
umns, temperature-compensated ECD, and mass fragmentography are prerequisites for successful analysis. All common amino acids can be separated in a single run on capillary
columns but the reproducibility can be improved for some of them, e.g., arginine. The
applications described in this chapter illustrate the feasibility of the micromethod in ecological
and medical research.

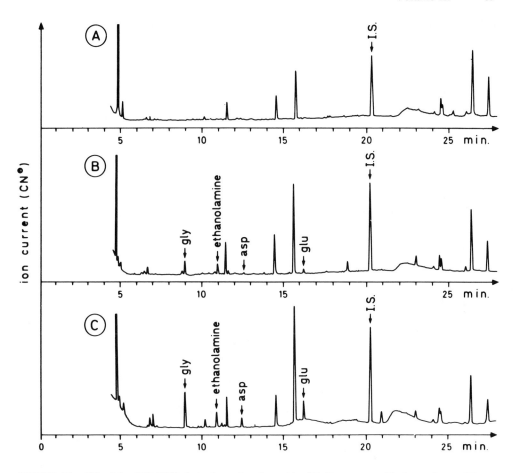

FIGURE 19.   GC of the PFP-HFIP derivatives of amino acids. (A) Represents a blank run from a Ringer-bicarbonate solution; (B) represents release of amino acids from the pigeon optic tectum during a resting situation; (C) obtained during electric stimulation of the optic nerve.[49]

REFERENCES

1. **Waller, G. R., Ed.,** *Biochemical Applications of Mass Spectrometry,* John Wiley & Sons, New York, 1972.
2. **Waller, G. R. and Dermer, O. C., Eds.,** *Biochemical Applications of Mass Spectrometry,* Vol. 1 (Suppl.), John Wiley & Sons, New York, 1980.
3. **Odham, G., Larsson, L., and Mårdh, P.-A., Eds.,** *Gas Chromatography/Mass Spectrometry — Applications in Microbiology,* Plenum Press, New York, 1983.
4. **Gardner, W. S. and Lee, G. F.,** Gas chromatographic procedure to analyze amino acids in lake waters, *Environ. Sci. Technol.,* 7, 719, 1973.
5. **Bengtsson, G. and Odham, G.,** A micromethod for the analysis of free amino acids by gas chromatography and its application to biological systems, *Anal. Biochem.,* 92, 426, 1979.
6. **Adams, R. F.,** Determination of amino acid profiles in biological samples by gas chromatography, *J. Chromatogr.,* 95, 189, 1974.
7. **Adams, R. F., Vandemark, F. L., and Schmidt, G. J.,** Ultramicro GC determination of amino acids using glass open tubular columns and a nitrogen-selective detector, *J. Chromatogr. Sci.,* 15, 63, 1977.
8. **Krutz, M.,** Untersuchungen zur Analyse von Aminosäuren in Oberflächenwasser, *Z. Anal. Chem.,* 273, 123, 1975.

9. **Cancalon, P. and Klingman, J. D.,** An improved procedure for preparing the *n*-butyl-trifluoroacetyl-amino acid derivatives and its application in the study of radioactive amino acids from biological sources, *J. Chromatogr. Sci.,* 12, 349, 1974.

10. **Hušek, P. and Macek, K.,** Gas chromatography of amino acids, *J. Chromatogr.,* 113, 139, 1975.

11. **March, J. F.,** A modified technique for the quantitative analysis of amino acids by gas chromatography using heptafluorobutyric *n*-propyl derivatives, *Anal. Biochem.,* 62, 420, 1975.

12. **Grob, K. and Grob, G.,** Techniques of capillary gas chromatography. Possibilities of the full utilization of high-performance columns. I. Direct sample injection, *Chromatographia,* 5, 3, 1972.

13. **Grob, K. and Grob, G.,** Splitless injection and the solvent effect, *HRC & CC,* 1, 57, 1978.

14. **Schomburg, G. and Husmann, H.,** Methods and techniques of gas chromatography with glass capillary columns, *Chromatographia,* 8, 517, 1975.

15. **Jennings, W. G., Freeman, R. R., and Rooney, T. A.,** A theoretical basis for the "solvent effect", *HRC & CC,* 1, 275, 1978.

16. **Schomburg, G., Behlan, H., Dielmann, R., Weeke, F., and Husmann, H.,** Sampling techniques in capillary gas chromatography, *J. Chromatogr.,* 142, 87, 1977.

17. **Schomburg, G., Husmann, H., and Rittmann, R.,** "Direct" (on-column) sampling into glass capillary columns. Comparative investigations on split, splitless and on-column sampling, *J. Chromatogr.,* 204, 85, 1981.

18. **Moodie, I. M. and Burger, J.,** Gas-liquid chromatography of amino acids: columns and methodology as a basis for routine amino acid analysis using glass capillary gas chromatography, *HRC & CC,* 4, 218, 1981.

19. **Ôi, N., Kitahara, H., and Doi, T.,** Gas chromatographic separation of amino acid, amine and carboxylic acid enantiomers with α-hydroxycarboxylic acid esters as chiral stationary phases, *J. Chromatogr.,* 207, 252, 1981.

20. **König, W. A., Parr, W., Lichtenstein, H. A., Bayer, E., and Oró, J.,** Gas chromatographic separation of amino acids and their enantiomers: non-polar stationary phases and a new optically active phase, *J. Chromatogr. Sci.,* 8, 183, 1970.

21. **Charles, R., Beitter, U., Feibush, B., and Gil-Av, E.,** Separation of enantiomers on packed columns containing optically active diamide phases, *J. Chromatogr.,* 112, 121, 1975.

22. **Frank, H., Nicholson, G., and Bayer, E.,** Rapid gas chromatographic separation of amino acid enantiomers with a novel chiral stationary phase, *J. Chromatogr. Sci.,* 15, 174, 1977.

23. **Saeed, T., Sandra, P., and Verzele, M.,** Separation of the enantiomers of proline and secondary amines, *HRC & CC,* 3, 35, 1980.

24. **König, W. A. and Benecke, I.,** Gas chromatographic separation of enantiomers of amines and amino alcohols on chiral stationary phases, *J. Chromatogr.,* 209, 91, 1981.

25. **Odham, G., Tunlid, A., Larsson, L., and Mårdh, P.-A.,** Mass spectrometric determination of selected microbial constituents using fused silica and chiral glass capillary gas chromatography, *Chromatographia,* 16, 83, 1982.

26. **Clarke, D. D., Wilk, S., and Ditlow, S.,** Electron capture properties of halogenated amine derivatives, *J. Gas Chromatogr.,* 4, 310, 1966.

27. **Bengtsson, G., Odham, G., and Westerdahl, G.,** Glass capillary gas-chromatographic analysis of free amino acids in biological microenvironments using electron capture or selected ion-monitoring detection, *Anal. Biochem.,* 111, 163, 1981.

28. **Pearce, R. J.,** Amino acid analysis by gas liquid chromatography of *n*-heptafluorobutyryl isobutyl esters. Complete resolution using a support-coated open-tubular capillary column, *J. Chromatogr.,* 136, 113, 1977.

29. **Abderhalden, E. and Weil, A.,** Über die bei der Isolierung der Monoaminosäuren mit Hilfe der Ester-methode entstehenden Verluste, *Hoppe Seyler's Z. Physiol. Chem.,* 74, 445, 1911.

30. **Bengtsson, G. and Cavallin, C.,** Compensation of baseline drift in temperature-programmed capillary gas chromatography with electron capture detection, *J. Chromatogr.,* 240, 488, 1982.

31. **Hites, R. A. and Bieman, K.,** Computer evaluation of continuously scanned mass spectra of gas chromatographic effluents, *Anal. Chem.,* 42, 855, 1971.

32. **MacKenzie, S. L. and Hogge, L.,** Gas chromatography mass spectrometry of the N(O)-heptafluorobutyryl isobutyl esters of the protein amino acids using electron impact ionisation, *J. Chromatogr.,* 132, 485, 1977.

33. **Tunlid, A. and Odham, G.,** Capillary gas chromatography using electron capture or selected ion monitoring detection for the determination of muramic acid, diaminopimelic acid and the ratio of D/L alanine in bacteria, *J. Microbiol. Methods,* 1, 63, 1983.

34. **Frank, H., Rettenmeier, A., Weicker, H., Nicholson, G., and Bayer, E.,** Determination of enantiomer-labelled amino acids in small volumes of blood by gas chromatography, *Anal. Chem.,* 54, 715, 1982.

35. **Raffer, J. J., Ingelman-Sundberg, M., and Gustafsson, J.-Å.,** Protein amino acid analysis by an isotope ratio gas chromatography-mass spectrometry-computer technique, *Biomed. Mass Spectrom.,* 6, 317, 1979.

36. **Coutts, R. T., Jones, G. R., and Liu, S. F.,** Quantitative gas chromatography/mass spectrometry of trace amounts of glutamic acid in water samples, *J. Chromatogr. Sci.,* 17, 551, 1979.

37. **Matthews, D. E., Ben-Galim, E., and Bier, D. M.,** Determination of stable isotopic enrichment in individual plasma amino acids by chemical ionization mass spectrometry, *Anal. Chem.,* 51, 80, 1979.
38. **Vetter, W.,** Amino acids, in *Biochemical Applications of Mass Spectrometry,* Vol. 1 (Suppl.), Waller, G. R. and Dermer, O. C., Eds., John Wiley & Sons, New York, 1980, chap. 14.
39. **MacKenzie, S. L.,** Amino acids and peptides, in *Gas Chromatography/Mass Spectrometry Applications in Microbiology,* Odham, G., Larsson, L., and Mårdh, P.-A., Eds., Plenum Press, New York, 1983, chap. 5.
40. **Rittenberg, D.,** The preparation of gas samples for mass-spectrographic isotope analysis, in *Preparation and Measurement of Isotopic Tracers,* Wilson, D. W., Nier, A. O. C., and Riemann, S. P., Eds., J. W. Edwards, Ann Arbor, Mich., 1946, 31.
41. **Sims, A. P. and Folkes, B. F.,** A kinetic study of the assimilation of (^{15}N)-ammonia and the synthesis of amino acids in an exponentially growing culture of *Candida utilis, Proc. R. Soc. London, Ser. B,* 159, 479, 1964.
42. **Rhodes, D., Myers, A. C., and Jamieson, G.,** Gas chromatography-mass spectrometry of N-heptafluorobutyryl isobutyl esters of amino acids in the analysis of the kinetics of [^{15}N]H$_4^+$ assimilation in *Lemna minor* L., *Plant. Physiol.,* 68, 1197, 1981.
43. **Robinson, J. R., Starratt, A. N., and Schlahetka, E. E.,** Estimation of nitrogen-15 levels in derivatized amino acids using gas chromatography quadrupole mass spectrometry with chemical ionization and selected ion monitoring, *Biomed. Mass Spectrom.,* 5, 648, 1978.
44. **Nordbring-Hertz, B.,** Peptide-induced morphogenesis in the nematode-trapping fungus *Arthrobotrys oligospora, Physiol. Plant.,* 29, 223, 1973.
45. **Odham, G., Larsson, L., and Mårdh, P.-A.,** Quantitative mass spectrometry and its application in microbiology, in *Gas Chromatography/Mass Spectrometry Applications in Microbiology,* Odham, G., Larsson, L., and Mårdh, P.-A., Eds., Plenum Press, New York, 1983, chap. 9.
46. **Larsson, L., Mårdh, P.-A., and Odham, G.,** Gas chromatography and mass spectrometry as diagnostic tools in clinical microbiology, *Lab. Manag.,* 21, 38, 1983.
47. **Bertilsson, L. and Costa, E.,** Mass fragmentographic quantitation of glutamic acid and γ-aminobutyric acid in cerebellar nuclei and symphathetic ganglia of rats, *J. Chromatogr.,* 118, 395, 1976.
48. **Wolfensberger, M., Reubi, J. C., Canzek, V., Redweik, U., Curtius, H. Ch., and Cuenod, M.,** Mass fragmentographic determination of endogenous glycine and glutamic acid released *in vivo* from the pigeon optic tectum. Effect of electric stimulation of a midbrain nucleus, *Brain Res.,* 224, 327, 1981.
49. **Wolfensberger, M., Amsler, U., Canzek, V., and Cuenod, M.,** Gas chromatographic method for the determination of trace amounts of putative amino acid neurotransmitters from brain perfusates collected *in vivo, J. Neurosci. Methods,* 5, 253, 1982.
50. **Odham, G. and Bengtsson, G.,** unpublished, 1983.

Chapter 5

CYCLIC AMINO ACID DERIVATIVES IN GAS CHROMATOGRAPHY

Petr Hušek

TABLE OF CONTENTS

I. INTRODUCTION

A. Alternatives to the Esterification Procedures

In the adjacent chapters it has been shown that methods, based upon derivatization of the α-amino group and esterification of the carboxyl group, in at least two steps, are powerful means for analysis of the protein amino acids quickly and precisely by gas chromatography (GC). However, even when the GC analysis of acylated amino acid alkyl esters became routine, some negative aspects of the esterification procedures persisted: (1) two incompatible reaction media with an intermediate evaporation step; (2) degradation of the amides, glutamine and asparagine, due to the HCl catalyst; (3) dissolution problems with some amino acids in the higher alcohols; (4) employment of high reaction temperatures for both reaction steps. There was, therefore, a lasting effort from the beginning to simplify the derivatization technique by introducing a universal reagent that would allow the reactive groups of the various amino acids to be treated simultaneously in one reaction step. The corresponding procedures are summarized in our review;[1] let us mention here briefly some of them.

First of all is the (trimethyl)silylation procedure, where by means of a potent silylating agent the amino acid is converted into a persilylated form (I). Even when the silylation approach has been developed into a sophisticated procedure that permits an effective GC analysis of all protein amino acids, the preparation of the derivatives, their extreme sensitivity to moisture (cleavage of the N–Si bond), and the long separation column required do not encourage one to choose this procedure for general use. However, a new silylation procedure with *N*-methyl-*N*-(*t*-butyldimethylsilyl)trifluoroacetamide (MTBSTFA) is under development at present, which is expected to be successful in converting of all protein amino acids into suitable derivatives for GC.[2] An opposite picture with regard to the derivative stability would be a procedure aimed at creation of peralkylated compounds, e.g., by treatment of amino acids with 2-bromopropane in the presence of sodium hydride. The derivatives (II) exhibit good GC properties but their preparation is far from quantitative and some multiple peaks are formed. Others

| I | II | III |

have tested dimethylformamide-dimethylacetal, which reacts with amino acids in the presence of acetonitrile to produce dimethylaminomethylene methyl esters (III). However, not all protein amino acids could be analyzed in this way. Finally, an attractive procedure employs a heat treatment with trifluoroacetic anhydride (TFAA) leading to formation of cyclic oxazolin-5-ones in a few minutes:

$$(1)$$

The simple amino acids gave good chromatographic peaks, unlike the multifunctional ones, which could not be determined in this way. To conclude, single-step derivatization techniques have generally presented difficulties for the quantitation of all protein amino acid members.

Having reviewed the single-step, single-reagent procedures, doubts remain as to whether "universal" chemical treatments are the best approach to deal with the protein amino acids and their polyfunctional nature. None of the approaches — except the questionable silylation procedure — has allowed analysis of all protein amino acids. Therefore, we have searched instead for a compromise, a reagent capable of derivatizing the α-amino and carboxyl groups simultaneously, and which allows a subsequent reaction step for other functional groups in the same milieu. We have found such a reagent among the perhalogenated acetones and by means of a "single medium, two-reagents" reaction course we are able to determine a complete spectrum of the protein amino acids and also the thyroid hormonal compounds, the iodothyronines and their analogs. The results of our investigation are summarized in this chapter.

B. Unique Features of the Halogenated Acetones

The reactions of the halogenoacetones were studied extensively by Simmons and Wiley.[3] The presence of the highly electronegative fluorine and/or chlorine atoms enhances the acidic character of the carbonyl group so that, in contrast to the behavior of aliphatic ketones, stable adducts with water, aliphatic alcohols, and ammonia are formed. In the case of 1,3-dichloro-1,1,3,3-tetrafluoroacetone (DCTFA), the equilibrium of the reaction

$$(CF_2Cl)_2C{=}O \ + \ H{-}X \ \rightleftharpoons \ (CF_2Cl)_2\overset{\underset{\displaystyle |}{OH}}{C}{-}X \qquad (\ X={-}OH,\ {-}OR,\ {-}NH_2\ ) \qquad (2)$$

is markedly shifted to the right. With α-substituted carboxylic acids, a stable cyclic derivative is formed, in which the two adjacent polar groups, the α-protonic (amino, hydroxy, or thiol) and the carboxylic, are coupled in a five-membered ring by a condensation reaction:

$$(3)$$

Substitution of X with $-NH$, $-O$, or $-S$ resulted in 2,2-bis(chlorodifluoromethyl)-4-subst.-1,3-oxazolidin-5-one (IV), -dioxolan-5-one (V), or -oxathiolan-5-one (VI), respectively.

IV   **V**   **VI**

The two chlorine and four fluorine atoms introduced in the molecule of the derivatives should contribute to an enhancement of the detection limit provided that the halogen-sensitive electron capture detector (ECD) were used. Moreover, the halogen atoms make the compound sufficiently volatile — as the parent agent is (DCTFA has a boiling point of 45°C) — which is advantageous for GC analysis. Last but not least, the reagent is relatively inexpensive, $\sim \$1.00$ per 1 or 2 mℓ (U.S. price at the end of 1970s) or per 1 g (European distributors in the 1980s), and not as poisonous as, for example, the perfluorinated anhydrides used for preparation of volatile derivatives for more than 20 years. The other halogenated acetone which could be considered, the hexafluoroacetone, is a gas or sesquihydrate, is more expensive, and does not contribute to the detector response to the same extent as DCTFA. Thus, the latter halogenoacetone only was taken for further experiments with amino acids.

C. Retrospection

To succeed in the condensation of α-substituted carboxylic acids with a halogenoacetone, a suitable solvent had to be found, as the acids themselves are not soluble in the reagent. Dimethylformamide, dimethylsulfoxide, and acetonitrile were used for this purpose in rather scarce studies[3-6] on condensation of DCTFA or hexafluoroacetone with some simple α-amino or α-hydroxy[6] carboxylic acids. Even in those potent organic solvents the conversion of amino acids to oxazolidinones occurred at elevated temperature only after tens of minutes or even several hours. Moreover, it has been reported that with the HCl salts of amino acids the reaction does not proceed.[4] Exceptions were the simple amino acid oxazolidinones, and also the condensed forms of the hydroxyl-containing amino acids which appeared on the chromatogram, either with free −OH groups or after silylation of −OH groups with hexamethyldisilazane.[4]

As is apparent, the results were not encouraging thus far. As luck would have it, our own investigation started without our knowing the earlier poor results. The optimal conditions for the condensation reaction of amino acids with DCTFA with respect to a solvent, catalyst, time and temperature, etc. were established using tyrosine and its mono- and diiodinated analogs as model compounds.[7,8] Various organic solvents were evaluated. In the presence of DCTFA the amino acids were best dissolved in dimethylsulfoxide (DMSO); the use of dimethylformamide required a little longer time. In acetonitrile, the amino acids were only partially dissolved at 50°C. However, GC analysis demonstrated clearly that in acetonitrile this dissolved portion was simultaneously converted to oxazolidinones. The addition of a small amount of pyridine to the acetonitrile (1:100 at least) improved the solubility of the amino acids remarkably, and the same results were obtained with amino acids and their hydrochlorides. Using this reaction medium the reaction time could be shortened from several hours to a few minutes. For the subsequent treatment of the side chain reactive groups, acylation with reactive anhydrides directly in the reaction medium proved to be the best method.

II. EXPERIMENTAL

A. GC Analysis

An ordinary dual-column instrument with heated injectors, temperature programmer, and two flame ionization detectors (FIDs) is required for the GC analysis of all protein amino acids as $N_1(O)$-acylated oxazolidinones. In our case a model 5730A Hewlett-Packard® gas chromatograph was used in combination with a 3380A computing integrator and later also with an 18740B capillary inlet system for special applications with capillary columns. Two glass columns of 2 mm I.D. and 2 m (column A) and 0.6 m (column B) in length were packed with fillings of the following composition:

Column A: 3% OV-17 (or SP-2250) methylphenylsilicone fluid on Chromosorb® W HP (or Supelcoport®), 80/100 mesh (a 3:1 blend of OV-17:OV-22 silicone phases was a better alternative for the Supelcoport®);
Column B: 1.5% SE-30 (or OV-1) methylsilicone gum on Chromosorb® G HP (80/100 mesh) or, alternatively, 3% SE-30 on Chromosorb® W AW DMCS (45/60 mesh).

The columns were operated under nitrogen (20 or 30 mℓ/min) in a temperature range of 75 to 230°C (8°/min increase) and with injector and detector temperatures of 200 and 250°C.

(The particular conditions are given in the legends to the figures.) Oxygen (60 mℓ/min) was found to afford a doubled response of the halogen-rich molecules when used for the hydrogen flame (30 mℓ/min) instead of air (240 mℓ/min) as a combustion supporter.[9]

B. Reagents and Materials

DCTFA was purchased from Fluka (Buchs, Switzerland) but has not been available from Fluka since 1983. As the same situation has occurred with other distributors, i.e., Aldrich-Europe (Beerse, Belgium), Riedel-de Haën (Seelze, West Germany), PCR (Gainesville, Fla.), Pfaltz & Bauer (Stanford, Conn.), etc., we would hope that a producer enters the market soon. For emergencies, there is a possibility to prepare the reagent by successive laboratory fluorination of 1,3-dichloroacetone, a very inexpensive compound (Fluka; Koch-Light, Colnbrook, Bucks, U.K.).

The anhydride reagents, heptafluorobutyric (HFBA), pentafluoropropionic (PFPA), and trifluoroacetic anhydride (TFAA), were purchased from Pierce Eurochemie (Rotterdam, The Netherlands). The organic solvents, acetonitrile, pyridine, benzene, petroleum ether (boiling range 60 to 80°C), hexane, heptane, dichloromethane, and methanol were from various sources in the best available grade. The chloroformates, methyl- (MCF), ethyl- (ECF), and isobutyl-chloroformate (IBCF) were obtained from E. Merck (Darmstadt, West Germany).

Amino acids of grade A quality were purchased from Calbiochem (Lucerne, Switzerland) or Sigma (St. Louis, Mo.). The protein amino acids, abbreviated in the figures as usual (except of cysteine and cystine, where the abbreviations CYSH and CYS are used), were included with some other amino acids, mainly to compile a series of homologs. Therefore, α-aminobutyric (ABA) and α-aminoisobutyric (AIBA) acids together with norvaline (NVal), norleucine (NLeu), and α-aminocaproic acid (ACA, first internal standard) were added to solutions of the carbon-linked chain amino acids; sarcosine and pipecolinic acid were added to solutions of the "imino" acids proline and hydroxyproline (Hyp); diaminobutyric acid (DABA), ornithine (Orn), and homoarginine (hArg) were added to basic amino acids; α-aminoadipic (AAA) and α-aminopimelic (APA) acids were added to the dicarboxylic amino acids, and diaminopimelic acid (DAPA, second internal standard), lanthionine, and homocystine (hCys) were added to cystine. For details see our earlier report.[10] Equimolar mixtures of particular amino acids were prepared in aqueous HCl (0.1 mol/ℓ) or in water (asparagine, glutamine, and tryptophan) to give 10 to 100 nmol of each amino acid in 10 μℓ of the aqueous medium. Approximately 1 M aqueous solutions of sodium carbonate (5 g/45 mℓ) and HCl (dilution 1:11) were prepared monthly from distilled water and chemicals of p.a. (Pro analysi) quality. Solvents for extraction (petroleum ether or hexane) were double-distilled, and DCTFA was treated with phosphorus pentoxide and distilled after some months of use.

The stock solutions of the reagents were kept in a refrigerator; work amounts (up to 1 mℓ) were placed in glass vials capped with Mininert® valves (Pierce) and kept at room temperature. The round-bottomed reaction vials (height approximately 50 mm, capacity approximately 2 mℓ) employed for derivatization and subsequent extractions were made from 12 mm O.D. glass tubes with ground glass joints 12.5/15 mm. Solid glass or bottom-closed stoppers were used. The reaction tubes were silanized by a treatment with dichlorodimethylsilane in toluene (1:10) for 30 min, and washed subsequently with methanol and acetone. After completed derivatization they can be reused after washing with acetone.

C. Analysis after Condensation

Add 40 μℓ of acetonitrile, 5 μℓ of pyridine, and 15 μℓ of DCTFA to dry residue of amino acids (less than 100 μg total) and allow to stand at room temperature for 5 min (for 10 min if the amount of proline is small). Then, an aliquot can be injected in the particular GC column.

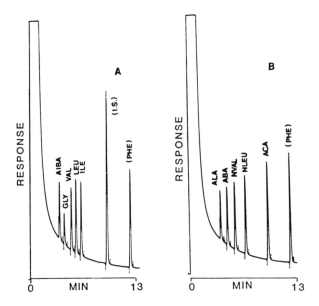

FIGURE 1.   GC analysis of aliphatic amino acids with (A) branched (plus glycine) and (B) straight carbon-linked chains on column A after 10 min of condensation. Phenylalanine (Phe) is added for comparison and *n*-hexadecane (I.S.) for evaluation of the absolute molar responses. Temperature range, 80 to 200°C.

Eleven protein amino acids can be readily analyzed in this way, and ten of them will elute from column A in the following order: alanine, glycine, valine, leucine, isoleucine, proline, methionine, phenylalanine, tyrosine (with free OH-group) and tryptophan (with free indolyl group, a partial elution only). Another column filling (column B) is required for elution of cystine and a full elution of tryptophan.

Results of GC analysis of aliphatic amino acids with two to eight carbon atoms in the molecule and a branched (A) or a straight (B) side chain structure are shown in Figure 1. It has been found that the side chain structure affects to a certain extent the course of the condensation reaction. The straight-chain ACA with eight carbon atoms and also Gly required a longer time for condensation, i.e., 5 to 10 min, to be cyclized fully (the same concerns cystine and the basic amino acids). Straight chain compounds with a smaller number of carbon atoms (Ala, ABA, NVal, and NLeu) and an amino acid branched in the γ-position (Leu) were converted into oxazolidinones within 3 to 5 min. Most quickly condensed were amino acids branched in the β-position (AIBA, Val, and Ile) and the hydroxyl- and sulfur-containing (Met, hCys) amino acids, within 1 to 2 min at 20°C.

Various results were achieved with condensation of "imino" acids, which should theoretically be more difficult because of the low reactivity of the "imino" hydrogen atom. However, *N*-methylglycine is in fact condensed to the oxazolidinone three to five times faster than is Gly. Likewise, condensation of pipecolinic acid proceeds in 1 to 2 min and the six-membered piperidine ring does not hinder the formation of the oxazolidinone ring (VII). In contrast, the oxazolidinone formation from the five-membered pyrrolidine of proline (VIII) is not a smooth process. The reaction course is influenced by medium polarity, reaction temperature, and the concentrations of pyridine and DCTFA in the medium. As the rate of condensation has an exponential course, about two thirds of proline is cyclized after 5 min under the given conditions. If proline is not present in the amino acid pattern, a lower amount of DCTFA and pyridine can be added to the acetonitrile in order to diminish the solvent tailing in the chromatographic column.

The OV-17 silicone phase was found to be convenient for the separation of Ala-Gly and Leu-Ile. Retention of branched chain amino acids in the column was less than that of the straight chain compounds. With AIBA the shift to lower retention times is so pronounced that it is eluted even before ALA. Concerning branching, it is interesting, however, that the γ-branched Leu is eluted before the β-branched Ile, contrary to expectation. On the other hand, elution of Gly behind Ala agrees with the results obtained previously with acylated amino acid alkyl esters.[1] This anomaly is probably caused by an enhanced interaction of the secondary amino group of the condensed Gly (IX) with the column packing, whereas with oxazolidinones of Ala (X) and sarcosine (N-methylglycine, XI) the presence of neighboring or with nitrogen-coupled methyl group decreases such an interaction. The retention times (r_t, relative to sarcosine) found confirm this conclusion:

IX	X	XI
$r_t = 1.17$	$r_t = 1.08$	$r_t = 1.00$

D. Analysis after Extraction

Add 400 μℓ of extraction medium (dichloromethane with petroleum ether, 1:3, or with hexane, 1:2, v/v) either alone or with 1 μℓ of chloroformate (MCF, ECF, or IBCF) to the sample after condensation and extract the derivatives into the organic phase by shaking the contents with 300 μℓ each of 1 *M* sodium carbonate, 1 *M* HCl, and water. A quick striking (5 to 7 times per second) of the tightly stoppered tube against a pad is effective in this respect and 10 to 15 sec are usually enough to clear the organic phase from a white pyridine precipitate (formed only if a chloroformate is present). The lower aqueous phase is always removed by means of a Pasteur pipet (e.g., joined to a microscrew on a fixed stand) and discarded. The upper layer is then transferred into another tube and evaporated as follows: (1) set the tube in such a place that the evaporation process can be followed visually (e.g., sand-bath heated not over 40°C); (2) evaporate the volatile self-cooling organic solvent to a small drop (20 to 30 μℓ) under a gentle stream of nitrogen: (3) evaporate the last drop by manually rotating the vial along its longitudinal axis; the dispersed drop can be dried more easily when, finally, 10 to 20 μℓ of extraction medium are added. It is necessary always to proceed according to item 3 when the simplest amino acids are present in the sample. Before GC analysis the derivatives are dissolved in heptane.

Isolation of the derivatives via extraction is necessary when the concentration of amino acids in the condensation medium is low, as pyridine tailing has an adverse effect on the analysis of the simplest members. Because of the subsequent evaporation the extraction medium has to be not only effective but also volatile, as explained. Considering these requirements, the combination of dichloromethane with light hydrocarbons (hexane is used as a substitute only when hydrocarbon impurities cannot be removed from the petroleum ether by distillation) proved to be best for this purpose. Direct evaporation of pyridine with the condensation medium always leads to complete loss of the oxazolidinones of most protein amino acids.

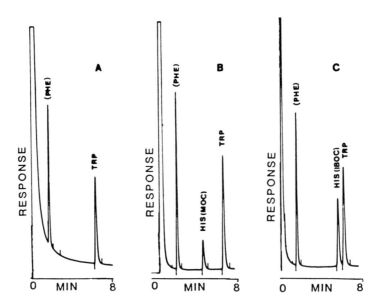

FIGURE 2.   GC analysis of tryptophan and histidine on column B after (A) condensation or condensation and extraction with added (B) methyl chloroformate or (C) isobutyl chloroformate. Temperature range, 120 to 200°C.

The aqueous acidic wash of the organic phase with the dilute HCl was found to be most effective in removing pyridine from the extract without deterioration of the derivative structure, even in the case of trifluoroacetylated hydroxyls in the amino acid side chain. The first alkaline wash, however, not only removes excess halogenated reagents (DCTFA and the reactive anhydrides used subsequently) and promotes extraction of the oxazolidinones into the organic phase by the salting-out effect of the nearly saturated sodium carbonate solution, but it has another important meaning with regard to histidine determination, which can be seen in Figure 2. While tryptophan can be eluted from the column B filling as a base, i.e., with a free indolyl group, the free imidazolyl group causes complete adsorption of histidine (as the oxazolidinone) in the column (Figure 2A). Moreover, if the extraction medium is not treated with a chloroformate, a cation is formed on the imidazole nitrogen during shaking with HCl and histidine passes into the aqueous phase.[10,11] Acylation with TFAA or HFBA does not lead to success, as the N^{im}-acyl moiety is hydrolyzed immediately during extraction and also the co-injection of the corresponding anhydride together with the sample in order to reacylate the imidazole does not allow elution of histidine oxazolidinone from the chromatographic column. Treatment of the organic layer with either of the three chloroformates has proved to be the best means of blocking the imidazolyl hydrogen effectively and for the formation of a stable derivative:

$$(CF_2Cl)_2 \quad NH \quad CH_2 \quad NH \quad + \quad Cl-COOR \quad \xrightarrow{Na_2CO_3} \quad (CF_2Cl)_2 \quad NH \quad CH_2 \quad N-COOR \tag{4}$$

The reaction proceeds instantaneously in the presence of small amounts of pyridine and carbonate in the reaction medium and the bulkiness of the attached radical influences not only retention times but also elution efficiency from the chromatographic column (Figure 2B and C). The elution of the N^{im}-methoxycarbonyl oxazolidinone of histidine was not complete when the alternative packing (support of W-type) for column B was used as it is in the figure. Much better results concerning the elution yield were obtained later with the

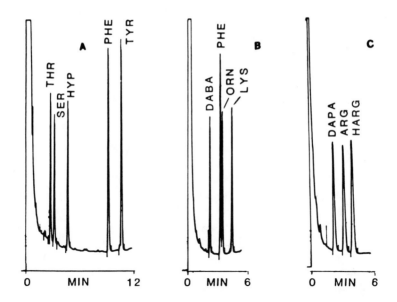

FIGURE 3.   GC analysis of (A) hydroxyamino acids, (B) α,ω-diamino acids and (C) arginine and homoarginine condensed and treated with HFBA according to procedure I (B,C) or II (A). After extraction and additional acylation the samples were analyzed on column A or B (sample C) in the temperature ranges 100 to 200°C (sample A) or 160 to 220°C (samples B and C).

G-type support, which is used at present in combination with the SE-30 silicone as the inherent packing of column B. From Figure 2 it is also apparent that the indolyl group of tryptophan was not attacked by the chloroformates.

E. Analysis after Acylation

Add 10 μℓ of HFBA (5 μℓ of TFAA or PFPA) either directly to the polar condensation medium (procedure I) or dilute the condensation medium, prior to anhydride addition, with 400 μℓ of the extraction medium (procedure II). After about 30 sec the extraction can be started (in the latter case, no more organic phase is added).

Additional acylation can be performed by adding a drop of the appropriate anhydride to the derivatives dissolved in heptane, and after heating the sample at 75 to 80°C for 2 to 3 min an aliquot is taken for GC analysis.

With the exception of tryptophan and also tyrosine, which could be analyzed even without acylation of its hydroxyl group as described earlier, the amino acids with side chain reactive groups require further chemical treatment. The reactive perfluorinated anhydrides proved to be most convenient for this purpose. An excess of DCTFA in the condensation medium promotes the dehydration process, so that acylation proceeeds instantaneously. Only citrulline, when present, requires a longer anhydride action according to procedure I (5 min at 40°C), during which the ureidic group is converted into an amino group so that this amino acid is then analyzed as ornithine.[10]

Oxazolidinones of amino acids with an additional amino or amido group are acylated effectively in a polar condensation medium (procedure I). Acylation of the α,ω-diamino acids, i.e., diaminobutyric acid, ornithine, and lysine is instantaneous and after extraction the N^ω-HFB (heptafluorobutyryl) oxazolidinones can be analyzed successfully in heptane only. However, application of the derivatives to column A in heptane with anhydride was preferred (Figure 3B), as the results were more reproducible. The use of TFAA instead of HFBA was also successful. Against expectations, the N^ω-TFA (trifluoroacetyl) oxazolidinones have higher retention times on column A than the HFB-analogs. The same is true for arginine and the *O*-acylated oxazolidinones with the OV-17 phase. The elution order changes when methylsilicones are used as stationary phases. In order to succeed in the determination of arginine, both the guanidine |–NH–C(=NH)–NH$_2$| terminal groups, i.e., the amino and the imino groups, must be acylated. Effective blocking of the guanidino group takes place after addition of anhydride to the condensation medium, so that the derivative passes into the organic phase. The acyl moiety on the imido group is nevertheless cleaved during this process, as analysis in heptane only produces no response. An additional acylation is, therefore, obligatory and the *N,N*-diHFB oxazolidinones of arginine and homoarginine can be eluted from column B completely (Figure 3C). When analyzed on column A, partial adsorption due to interaction with the chromatographic support occurs (arginine response diminished to about two thirds). In order to acylate the hydroxylated amino acids successfully, it was necessary to dilute the condensation medium with the extraction medium prior to anhydride addition (procedure II). Without the dilution the acylation yields of especially threonine and serine were low and threonine gave two peaks, the second being co-eluted with hydroxyproline. The acylation yields bear a close relationship to the dilution of the condensation medium up to a given volume and no further changes occur with additional dilution. Acylation of tyrosine was not influenced by the polarity of the medium to the same extent and direct acylation in the condensation medium gave a molar response lower by about one quarter. Condensation of hydroxyproline was approximately five times more rapid as that of proline because of an electron shift on the pyrrolidine ring due to the hydroxyl group, acylation of which according to procedure II proceeds smoothly. Analysis of *O*-HFB oxazolidinones of hydroxylated amino acids is shown in Figure 3A.

If the sample contains amino acids with thiol groups (cysteine, penicillamine) the medium immediately turns yellow after addition of DCTFA. The subsequent acylation does not alter the fact that the chromatographic analysis affords no peaks. Thus, in order to succeed with analysis of cysteine, the thiol group must be blocked in advance, e.g., by methylation to form *S*-methylcysteine or by treatment with formaldehyde to create thiazolidine-4-carboxylic acid.[10] Both forms are condensed readily and analyzed smoothly.

Acylation of the two hydroxyl groups in the molecule of dihydroxyphenylalanine and the one hydroxyl and one amino group in the molecule of hydroxylysine gave mostly unsatisfactory results with either of the two acylation procedures. Analysis of hydroxylysine gave only some minor peaks even after the additional acylation, that of dihydroxyphenylalanine afforded only about 50% of the expected response. The diHFB acylated oxazolidinone of dihydroxyphenylalanine is eluted just before tyrosine on column A.

F. Analysis after a Special Treatment

In GC of amino acids there is a lack of a convenient procedure enabling asparagine and glutamine to be estimated separately from their corresponding acids, as the most widely used methods based on esterification of the carboxyl group with acidified alcohols lead to conversion of the amides to the acids. Condensation of amino acids with DCTFA proceeds in a mild, weakly basic medium which does not attack the amido groups of asparagine and glutamine (Asn and Gln). However, there arises the problem of how to treat the second carboxyl group in molecules of aspartic (Asp) and glutamic (Glu) acids. Addition of an anhydride in the condensation medium (acylation procedure I) leads to unsatisfactory results because of the multiple-peak formation for both the dicarboxylic acids and glutamine (Figure 4). Only asparagine gives a full response as a single peak.

It seemed reasonable to find if there were a process that would allow esterification of the carboxyl group in the alkaline aprotic medium. Brooks and co-workers[12] reported on the simultaneous esterification of carboxyl and acylation of hydroxyl groups with alcohol and HFBA for the analysis of hydroxy acids by GC. A reagent mixture consisting of HFBA,

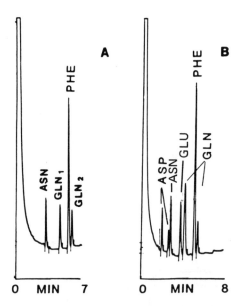

FIGURE 4.   GC analysis of equimolar mixture of Phe (the highest peak) with (A) Asn and Gln alone, (B) in mixture with Asp and Glu, treated with 10 μℓ HFBA after the performed condensation. After extraction and additive acylation in heptane two peaks appeared for each of the following amino acids: Asp (retention time 2.70 and 3.47 min), Glu (4.85 and 5.36 min), and Gln (5.39 and 6.67 min). Asn gave a single peak only (3.70 min).

pyridine, and ethanol was used to esterify carboxyl groups with the alcohol and to derivatize hydroxyl and amine groups with an excess of the anhydride. HFBA catalyzed esterification of carboxyl groups and pyridine catalyzed derivatization of carboxyl, hydroxyl, and amine groups. Good results were obtained in a few minutes without heating the sample.

We examined this approach in connection with our problem. In order to obtain more definitive results, the two higher homologs with six and seven carbon atoms (AAA and APA) were treated together with Asp and Glu. In accordance with the unique features of fluoroketones, any addition of alcohol results in immediate formation of alcohol-DCTFA adduct, so that with methanol, ethanol, and isopropanol the corresponding 2-methoxy- (XII), 2-ethoxy- (XIII), and 2-isopropyloxy- (XIV) DCTF-propane-2-ols are formed:

$$
\begin{array}{ccc}
CF_2Cl & CF_2Cl & CF_2Cl \\
| & | & | \\
HO-C-OCH_3 & HO-C-OCH_2CH_3 & HO-C-OCH(CH_3)_2 \\
| & | & | \\
CF_2Cl & CF_2Cl & CF_2Cl \\
\\
\textbf{XII} & \textbf{XIII} & \textbf{XIV}
\end{array}
$$

The corresponding halogenated alkoxy-alcohol does esterify the second carboxyl group in the presence of pyridine and added anhydride:

$$
\textbf{(ASP,GLU)} \quad R-COOH + HO-\underset{\underset{CF_2Cl}{|}}{\overset{\overset{CF_2Cl}{|}}{C}}-OR' \longrightarrow R-COO-\underset{\underset{CF_2Cl}{|}}{\overset{\overset{CF_2Cl}{|}}{C}}-OR' \qquad (5)
$$

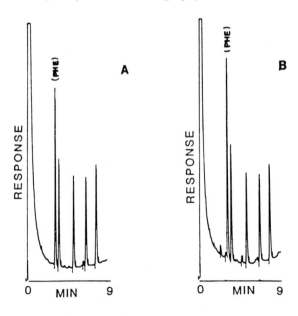

FIGURE 5.   GC analysis of equimolar mixture of Phe (first peak) with C_4 to C_7 dicarboxylic amino acids after condensation (30 $\mu\ell$ acetonitrile, 6 $\mu\ell$ pyridine, 15 $\mu\ell$ DCTFA), addition of benzene and methanol (50 + 3 $\mu\ell$), acylation with (A) 9 $\mu\ell$ HFBA or (B) 5 $\mu\ell$ TFAA, extraction and dissolution of the evaporated residue in heptane. The reverse elution order of Phe and Asp is due to change in packing (3% OV-101 on Supelcoport®) in a 1.5-m column operated in the range 150 to 230°C.

Figures 5 and 6 show results of GC analysis of the C_4 to C_7 dicarboxylic amino acids condensed to oxazolidinones and treated in the condensation medium with methanol and HFBA or TFAA (Figure 5A and B) or with ethanol or isopropanol and HFBA (Figure 6A and B). After extraction the evaporated residue was dissolved in heptane only and injected. Even from the figures it is apparent that the retention times are not dependent on the anhydride but on the alcohol used. The results are slightly higher and more consistent with HFBA, whereas with TFAA small interfering peaks occur on the chromatogram (Figure 5B). Methanol proved to be the alcohol of choice as with the higher alcohols the molar responses of the higher homologs decline. The structures of the derivatives were confirmed by MS.[13]

The optimal reaction conditions resulting from experimental studies are given in the legend to Figure 5. Benzene is added together with methanol to the condensation medium as the polarity of this medium at the moment of anhydride addition influences the esterification process to a certain extent; maximum yields are obtained with a benzene-acetonitrile volume ratio of 3:2. Even more important is the ratio between methanol, pyridine, and anhydride, being 1:1:0.5 by moles and, for HFBA, 1:2:3 by volumes for maximal reaction yields.[14] The GC analysis succeeds in heptane only and no additional treatment with HFBA is required.

For a simultaneous analysis of Asp and Glu with their amides, however, the additional acylation is necessary; otherwise Asn and Gln do not appear on the chromatogram from a packed column. Derivatization of Asn only was found to be influenced by both the total amount of anhydride added into the first reaction medium and time of the second acylation in heptane (Figure 7).[15] An additional portion of HFBA anhydride must be added for full response of Asn in order to balance the disturbing effect of the alcohol addition. To achieve the same maximal response as that without methanol addition a doubled portion of HFBA was added following the first one. A short time interval, 2 to 3 sec usually (or arbitrarily

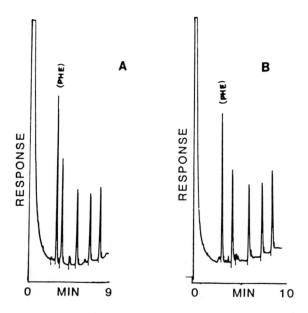

FIGURE 6. GC analysis of the same mixture of amino acids as in Figure 5 treated equally with only one exception: (A) 4.3 µℓ ethanol or (B) 5.6 µℓ isopropanol were used instead of methanol and HFBA was the only acylation reagent. The corresponding derivatives exhibit proportionally longer retention times.

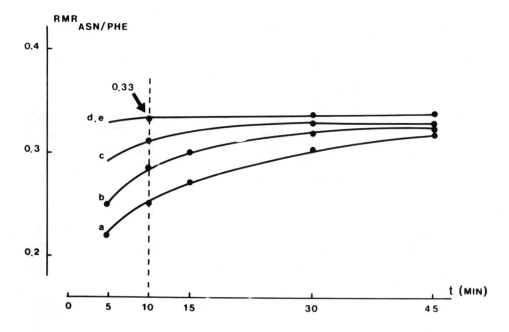

FIGURE 7. Effect of amount of HFBA added to the sample during first acylation and the time of the additional acylation on RMR (related to Phe) of Asn. The following amount of HFBA was added to the sample treated as in Figure 8: (a) 9 µℓ, (b) 9 + 9 µℓ, (c) 9 + 14 µℓ, (d) 9 + 18 µℓ. In (e) 10 µℓ of HFBA were added to the medium without alcohol addition.

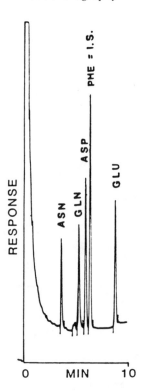

FIGURE 8.   GC analysis of dicarboxylic amino acids and their amides in an equimolar mixture with Phe on the same column as in Figure 4. The compounds were condensed at 20°C for 5 min in the same medium as in Figure 5 and after treatment with 70 μℓ of benzene-methanol mixture (22:1) two portions of HFBA (9 + 18 μℓ) were added. After extraction the additional acylation was performed at 75°C for 10 min.

longer), between the two additions was enough to keep the response of Asp and Glu without change. From Figure 7 it can also be seen that with addition of only the first HFBA portion the response of Asn declined to about 75%, which is acceptable in most cases. Gln does not require a second anhydride addition, with the presence of alcohol in the medium of lowered polarity allowing formation of a single peak only. GC analysis of dicarboxylic amino acids and their amides is shown in Figure 8. In the legend to the figure, the given reaction conditions show, when compared with those being optimal for the dicarboxylic amino acids, that a further dilution of methanol by benzene is the means to promote a response enhancement for ASN. This change in reaction conditions does not deteriorate analysis of the other members.

III. PREPARATION AND ANALYSIS OF PROTEIN AMINO ACID OXAZOLIDINONES

A. Procedure with Commentary
   Condensation — Add 50 μℓ of solvent (acetonitrile-pyridine, 5:1) and 20 μℓ DCTFA to the dry residue of amino acids (less than 0.1 to 0.2 mg total) and hold the sample at room temperature for 15 min.
   Acylation — Add 70 μℓ of acylation medium (benzene-methanol, 22:1) and two 10-μℓ

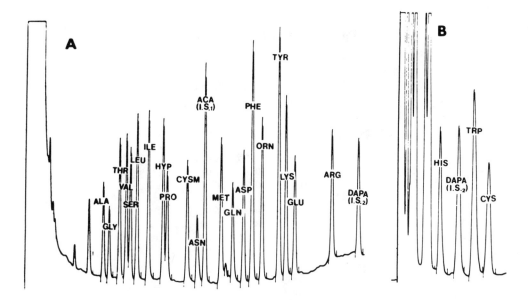

FIGURE 9.   GC analysis of (N,O)-HFB amino acid oxazolidinones after derivatization of an equimolar mixture containing 50 nmol of each amino acid. (The brackets around N,O mean that only the reactive side chain groups are acylated.) Temperature ranges: 75 to 230°C for column A and 150 to 200°C for column B. Each peak represents 1 nmol of amino acids (at 4×10^{-4} A).

volumes of HFBA (approximate time interval, 1 to 2 sec) to the condensation medium with continuous gentle mixing and allow to react for at least 30 sec.

Extraction — Add 400 $\mu\ell$ of extraction medium (petroleum ether or hexane-dichloromethane-methyl chloroformate, 200:100:1) to the sample and proceed further as described in Section II.D. If a drop of water appears in the tube after evaporation of the organic phase it should be removed before the heptane addition by contacting it with a piece of blotting paper.

Additive acylation — Add 20 to 50 $\mu\ell$ of heptane to the residue of the derivatized amino acids and inject an aliquot on column B. Then add a drop (2 to 3 $\mu\ell$) of HFBA to the sample and after heating at 80°C for 3 min inject an aliquot on column A.

The described procedure[16] is a result of the preceding attempts to find a chemical treatment that would deal successfully with all of the reactive side chain groups in a mixture of the protein amino acids. As the rate of dissolution of the compounds in the condensation solvent plays an important role, a limited amount of material should be treated as stated. The pyridine concentration influences the derivative yields for some amino acids and by enhancement of its amount in the solvent, the dissolution of the amino acid residue and cyclization of the two imino acids is more rapid. The selected reaction conditions and condensation time of 15 min represent a compromise in respect of the cyclization of proline; the formation of the cyclic form (retention time 8.56 min, see Figure 9) was stimulated even when a small portion remained uncyclized and underwent a second reaction with HFBA to give a side product (small peak behind methionine with a retention time of 11.96 min). Provided the amino acid sample lacks proline, a condensation time of 5 min is sufficient.

An acylation time of at least 30 sec is required for a full response of glutamine. The two-step HFBA addition improves the results for dicarboxylic amino acids, and the total volume of 20 $\mu\ell$ proved to be sufficient for an effective treatment of asparagine and arginine. Under the chosen acylation conditions the indolyl group of tryptophan is not acylated.

Optimization of the conditions led to a lowering of the amount of methanol added to about 75% of that previously recommended.[14] This was done, together with enhancement

Table 1

MOLAR RESPONSES OF AMINO ACID OXAZOLIDINONES RELATIVE TO α-AMINOCAPRYLIC ACID OR TO DIAMINOPIMELIC ACID (HIS, TRP AND CYS ONLY)

Amino acid	Abbr.	$\bar{x}$	SD	CV (%)
Alanine	Ala	0.41	0.016	3.93
Glycine	Gly	0.30	0.016	5.25
Threonine	Thr	0.65	0.018	2.72
Valine	Val	0.64	0.029	4.57
Serine	Ser	0.60	0.025	4.16
Leucine	Leu	0.76	0.025	3.28
Isoleucine	Ile	0.78	0.008	1.06
Hydroxyproline	Hyp	0.73	0.041	5.75
Proline	Pro	0.54	0.020	3.63
S-Methyl cysteine	Cysm	0.59	0.012	2.05
Asparagine	Asn	0.36	0.031	8.61
Methionine	Met	0.70	0.030	4.24
Glutamine	Gln	0.43	0.039	8.83
Aspartic acid	Asp	0.65	0.034	5.18
Phenylalanine	Phe	1.12	0.008	0.73
Ornithine	Orn	0.76	0.031	4.08
Tyrosine	Tyr	1.17	0.032	2.73
Lysine	Lys	0.84	0.029	3.44
Glutamic acid	Glu	0.59	0.020	3.38
Arginine	Arg	0.72	0.047	6.55
Histidine	His	0.80	0.038	4.75
Tryptophan	Trp	1.52	0.031	2.04
Cystine	Cys	0.84	0.029	3.44

Note: Values shown are the means for ten individually prepared samples of an amino acid calibration mixture (total amino content less than 0.2 mg) together with standard deviations (SD) and coefficients of variation (CV). The amino acids are ordered according to their elution from the analytical column(s).

of the amount of acetonitrile in the milieu, in order to get a full response for arginine. Under this compromise the yields were about 10% lower for the dicarboxylic amino acids, 20% lower for asparagine, and 5 to 7% lower for lysine and ornithine than the maximal responses. The relative molar responses (RMR) including the coefficients of variation (CV) are given in Table 1. For comparison with the absolute molar response values see Reference 9. The overall reproducibility averages 4%, the worst cases being glutamine and asparagine (8 to 9%) and also arginine (6 to 7%).

The small amount of methyl chloroformate (we prefer use of ethyl chloroformate at present) in the extraction medium proved to be an effective means for converting the imidazolyl group in the histidine oxazolidinone to a stable analyzable form instantaneously during shaking of the organic phase with the first alkaline wash. The free indolyl group of tryptophan as well as the other derivatives are not attacked.

Three amino acids, i.e., arginine, glutamine, and asparagine, cannot be analyzed without additional acylation in heptane at a higher temperature, during which the guanidino group of arginine is acylated fully and the amides are converted into the nitriles. As any change in the asparagine response occurred when the heating was prolonged from 3 to 10 min, the

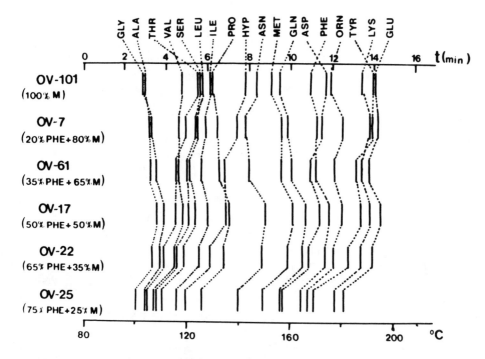

FIGURE 10. Retention behavior of (*N*,*O*)-HFB oxazolidinones of 18 amino acids in a column (2 m × 2 mm) packed with 3% of the particular phase on Chromosorb® W HP (80/100 mesh). Temperature range: 80 to 240°C (8°/min). Carrier gas flow rate: 25 mℓ/min. PHE = phenyl, M = methyl.

shorter heat treatment is used in general. As tryptophan undergoes a partial and successive acylation on the indolyl hydrogen and the response of histidine declines a little due to the heat treatment of the additive acylation, a preliminary analysis on column B in heptane only is therefore preferred. The other derivatized amino acids can be analyzed on column A also in heptane only but a partial adsorption (10 to 20%) of the *N*- and *O*-HFB acylated forms takes place.

B. Analysis by Means of Two Packed Columns

The separation of the protein amino acids, along with hydroxyproline, *S*-methylcysteine, and ornithine and two internal standards is shown in Figure 9. Chromosorb® G was found to be an excellent chromatographic support for complete elution of histidine, tryptophan, and cystine oxazolidinones from the shorter B column. The other protein amino acid derivatives were best separated on a 2-m column with a milder polar silicone phase. Chromosorb® W HP proved to be the best of the three high-quality diatomaceous supports tested (plus Supelcoport® and Gas-Chrom® Q), as mere coating with the OV-17 phase afforded a satisfactory resolution of the pair proline-hydroxyproline. If Supelcoport® is used instead of Chromosorb®, an enhancement of the phase polarity (addition of OV-22) is required as stated in the experimental. Broadening of peaks occurred when Gas-Chrom® Q was used.

The influence of phase polarity on the retention behavior of the derivatives is shown in Figure 10. Even on the least polar methylsilicone phase (OV-101) the pair leucine-isoleucine is separated completely, while threonine-serine (together with leucine) and some other amino acids are coeluted. As the phenyl group content of the phase increases, the separation of alanine-glycine and threonine-serine pairs improves and all the hydroxyl-containing amino acids, together with aspartic acid, exhibit an almost linear decrease in retention times. The best separation of all the compounds is achieved on the OV-17 phase; an improved separation of proline-hydroxyproline is possible by mixing with OV-22, as recommended for Supelcoport®.

The retention behavior of threonine and serine in the analytical column A is influenced by the carrier gas flow rate and initial working temperature: higher flow rates result in higher retention times relative to valine for these two compounds; a higher initial temperature leads to the opposite behavior. However, the separation of the two imino acids is worse under such conditions, as hydroxyproline exhibits higher retention times and tends to merge with proline.

Instead of HFBA, TFAA (or PFPA) can be used as the acylation agent. The TFAA-treated oxazolidinones have lower retention times only on the phases with lower contents of phenyl groups, while on OV-17 these derivatives are eluted later than the corresponding HFBA-treated forms.[10] None of the phases studied enabled a complete separation of the TFAA-treated oxazolidinones, so that the employment of HFBA was generally preferred.

C. Application to Biological Fluids

For routine amino acid analysis in minute amounts of physiological fluids a pretreatment of the starting material, often involving a cleanup isolation step on ion exchangers, is carried out prior to derivatization. According to our experiences, a method involving dilution of the sample with acetic acid and thus eliminating the protein precipitation, affords more reproducible results. This fact led us to investigate the acetic acid procedure more comprehensively, i.e., with respect to the ion exchange material and the inherent isolation process. It was found with the strongly acidic cation exchangers of the Dowex® 50W type that lowering of the percentage of cross-linking from 8 to 2% resulted in better recoveries for the aromatic and long chain aliphatic amino acids and, that by using reversed flow for sample application a further improvement occurred.[17] A recommended procedure for isolation of free amino acids from 50 to 100 $\mu\ell$ of serum or urine is the following.

Thin-wall polyethylene tubing (3-3.2 mm I.D.), commonly available, was cut into pieces about 25 mm long, and a plug of glass wool (3 to 4 mm) was placed in one end. A slurry of 100/200 mesh Dowex® 50W-X2 (H$^+$) p.a. resin in water was sucked in the tubing with help of a syringe to form a 15-mm high column of the wet resin. The resin bed was washed subsequently with 1 mℓ of 1 M HCl and 2 mℓ of distilled water.

A 50- to 100-$\mu\ell$ volume of serum (or urine) was placed in a 2-mℓ vial and mixed with ten times the volume of 25% (v/v) aqueous acetic acid, containing internal standards (25 nmol/mℓ). The mixed solution was sucked slowly (about 0.5 mℓ/min) through the activated resin in the plastic column, the lower end of which was immersed in the liquid, with the upper end connected by a piece of glass stem (3.2 mm O.D. and 2 mm I.D.) to the silicone tubing of a peristaltic pump. After the fluid had been completely withdrawn from the vial, 250 to 500 $\mu\ell$ of water were added to the vial and sucked into the resin at a flow rate increased by two- to threefold. This step was repeated. For elution of the amino acids it was necessary to place in the lower ''suck-in'' end of the column a piece of polypropylene tip (a conventional conical tip, used for push-button pipettes, cut on both ends to about 3.2 and 2 mm O.D., respectively) filled with glass wool. The upper end of the column was then connected to the PTFE luer (8 mm × 3.2 mm O.D.) of a 1-mℓ syringe, which was just filled with 0.7 mℓ of 2 M aqueous ammonia, and the amino acids were eluted into a vial by pushing the syringe plunger down for 15 to 20 sec.

Examples of analyses of free amino acids in human serum and urine of control subjects on column A are shown in Figure 11.

D. Single Column Analysis

The necessity of using a second column for the elution of histidine, tryptophan, and cystine oxazolidinones is an apparent drawback of this derivatization approach. The perhalogenated

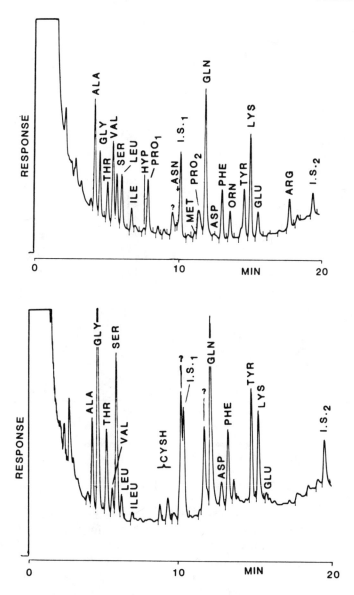

FIGURE 11.   Analysis of free amino acids in human control serum (above) and urine (below) after work-up of 100 µℓ of the corresponding fluid according to the procedure in the text on column (A). Internal standards added in amounts of 10 nmol.

derivatives are more prone to adsorption than any other derivatives used, and the chromatographic support is responsible for the adsorption, Chromosorb® 750 has been recommended as an inert, efficient support of high quality. We have tested it in combination with a blend of methylphenylsilicone phases as packing for column A (Figure 12). Even when a partial elution of histidine, tryptophan, and cystine was possible from this packing at a high flow rate of carrier hydrogen, the separation efficiency of the better deactivated Chromosorb® 750 support is less than that of the W type. Furthermore, as the HFBA-treated oxazolidinones could not be separated under the selected conditions, treatment with PFPA (8 + 8 µℓ PFPA were used for the acylation) was chosen instead of HFBA. This was a last alternative, at present, toward separation of all 20 protein amino acids as oxazolidinones on a single packed column.

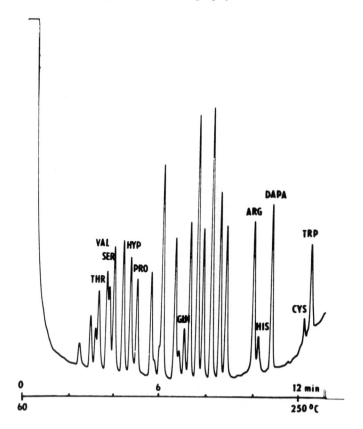

FIGURE 12.   PFPA-treated oxazolidinones analyzed on a 2 m × 2 mm
glass column packed with 2% OV-17-OV-22 (1:1) on Chromosorb® 750
(80/100 mesh) operated in the temperature range 60 to 250°C with a
hydrogen carrier gas flow of 60 mℓ/min. For assignment of amino acids
to corresponding peaks see Figure 9.

A search for a chromatographic system with higher performance with respect to single-
column analysis of the protein amino acids led us to investigate open tubular columns. Our
primary aim was not to increase the resolution power or to accelerate the analysis, but
explicitly to exclude the chromatographic support as the predictable source of adsorption.
Simultaneously, we wished to retain the possibility for direct sample injection of larger
volumes using conventional injection systems, and for this reason we focused on wide-bore
capillary columns. Attention was centered on capillaries with methylphenylsilicone phases,
because of good experiences gained with them. Because of the known difficulties in coating
fused-silica open tubular (FSOT) columns with polar silicone phases such as OV-17, we
tried the recommended substitute OV-1701 (chemically bonded), a silicone gum similar in
polarity to OV-17 liquid. However, because of the different composition of the phase
(cyanopropyl groups introduced) the elution order changed, the separation was unsatisfactory,
and the heaviest members were not eluted.

We then turned our attention to 0.5 mm I.D. and 12 m long glass wall-coated open tubular
(WCOT) column with a 0.45-μm layer thickness of OV-17 phase (Packard-Becker, Delft,
The Netherlands). A direct sample injection via splitless mode into a heated (200°C) 9 cm
× 2 mm I.D. glass insert (injected amount, 1 μℓ) led to successful analysis of all protein
amino acids in 10 min (16°C per minute temperature increase)[18] or even in 5 min (Figure
13). Among all known techniques and GC approaches this is a shortest time for obtaining
a complete amino acid spectrum.

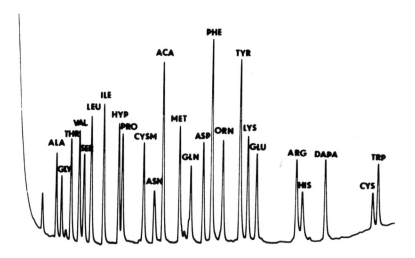

FIGURE 13. HFBA-treated oxazolidinones analyzed on a 12 m × 0.5 mm glass WCOT (OV-17) column in the temperature range 70 to 240°C (32°/min temperature increase) with a hydrogen flow of 9 mℓ/min.

A satisfactory separation was even achieved with the widely used methylsilicone phases, coated in capillaries of 25 m length (Figure 14). Again, PFPA-treated oxazolidinones were a better choice as with the HFBA-derivatives some peaks were coeluted. Comparing both chromatograms on the figure, it is obvious that the analytical performance of the FSOT column with bonded methylsilicone gum is superior to that of the WCOT column. Both the amides, Asn and Gln, fail on the chromatogram of the latter column, confirming the presence of residual adsorption in this column. Unfortunately, Asn is the only amino acid which cannot be separated from another one, Ile, on the FSOT column.

Possibilities for analysis in the subpicomole range with the OV-1 cross-linked FSOT column were tested after connecting it to an electron capture detector. Whereas with the packed column A adsorption losses in the column fillings occur with the side chain acylated amino acids when the injected amount is lower than 100 pmol per injected compound (the most drastic decline was observed with the hydroxyl-containing amino acids, especially serine), with the OV-1 FSOT column no adsorptive losses were observed till 0.01 pmol of injected mass and no changes in RMR of the protein members appeared[19] (Figure 15). This means, therefore, that employment of open tubular columns without a chromatographic support is the only means of obviating losses of the perfluoroacylated amino acid oxazolidinones at concentration ranges below 100 pmol per injected amino acid derivative. Also it is worth noting that the presence of hydrogen, being the carrier gas for the capillary column, did not seem to influence the process of electron capture in the detector provided that nitrogen was added as the make-up gas. The results show that the GC-ECD analysis of perfluoroacylated amino acid oxazolidinones near and below the picomole range is very possible; however, one has to face impurities of various origins and types when the initial derivatized amount is low.[19]

IV. FAST PROCEDURE FOR GC ANALYSIS OF THYROID HORMONAL COMPOUNDS

In the 1980s there is an apparent trend toward the use of GC for the analysis of thyroid hormones in the form of *N,O*-HFB methyl esters. The drawback of the esterification-acylation procedure is the long derivatization time of about 2 hr and the necessity of a fresh 25%

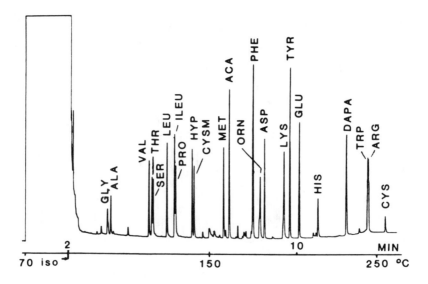

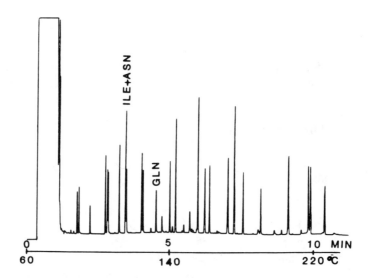

FIGURE 14. PFPA-treated oxazolidinones analysed on 25-m long capillaries coated with OV-101 methylsilicone fluid in glass (0.25 mm I.D., above) or with OV-1 methylsilicone gum in fused silica (cross-linked, 0.31 mm I.D., below). The corresponding temperature ranges were 70 (hold for 2 min) to 250°C and 60 to 230°C (16°/min increase in both cases) and the hydrogen flow rates 2.9 and 4.6 mℓ/min, respectively.

gaseous HCl-methanol solution. Based on condensation with DCTFA we describe here, indeed, a very fast procedure that is suitable not only for the iodoamino acids, diiodothyronine (T_2), triiodothyronine (T_3), reverse triiodothyronine (rT_3), and thyroxine (T_4), but also for their deamination and decarboxylation products, i.e., for triiodothyroacetic acid (T_3Ac) and tetraiodothyroacetic acid (T_4Ac). The following treatment was performed: in a 2-mℓ glass reaction vial, the dry residue of the compounds of interest was covered with 120 µℓ of acetonitrile-pyridine (40:1, v/v) and 15 µℓ DCTFA, and after about 1 to 2 min 6 µℓ HFBA were added. After a further 1 to 2 min the derivatized forms were extracted into hexane (400 µℓ) by shaking with water (300 µℓ) for about 10 sec. The lower phase was sucked

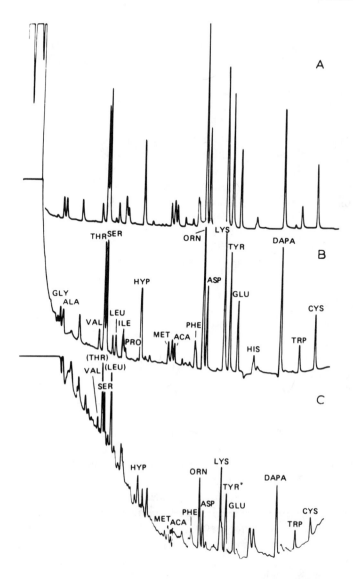

FIGURE 15.   GC-ECD analysis of (*N*,*O*)-PFP amino acid oxazolidinones in 25 m × 0.31 mm FSOT (OV-1 cross-linked) column in 60 to 230°C temperature range (H$_2$ flow: 4.6 mℓ/min, N$_2$ make-up flow: 20 mℓ/min). An equimolar mixture with 5 nmol of each amino acid was derivatized and the sample was diluted to the following final amounts injected: (A) 1 pmol (attenuation × 32), (B) 0.1 pmol (× 4), (C) 0.01 pmol (× 1). Amino acids in brackets are coeluted with an unknown impurity having identical retention time.

off and the organic solvent evaporated at 50 to 60°C under a stream of nitrogen in the same or a second reaction vial. After addition of heptane (20 to 50 μℓ) the sample was subjected to GC. The analysis was done on a 5 m × 0.31 mm I.D. FSOT column with OV-1 cross-linked phase (layer thickness 0.17 μm). The column was operated in the range of 190°C (hold for 2 min) to 240°C with a linear temperature increase of 4°/min and a hydrogen flow rate of 4.7 mℓ/min. The nitrogen make-up flow rate was equal to that of additional hydrogen for the FID (300°C), i.e., 30 mℓ/min. The splitless mode injection technique was used.[20] As the internal standard (I.S.), 2-methylnaphthalene-bis(hexachlorocyclopentadiene) adduct from Aldrich Europe (Beerse, Belgium) was used.[21] For GC analysis see Figure 16.

Under the stated conditions of the condensation medium the ring closure need not be complete as the subsequent addition of the reactive anhydride results in a cooperative effect. The same is true for acylation of the phenolic group, where excess of DCTFA promotes the dehydration process, which is then immediate. Thus, the cooperative action of both strong dehydrating agents enables the whole procedure to be shortened to a few minutes. The molar ratio between HFBA and pyridine is important concerning the yields for T_3Ac and T_4Ac. A slight molar excess of the anhydride results in acidification of the aqueous phase during extraction and pyridine is removed from hexane in this way.

The derivatized compounds are really perhalogenated substances of high molecular weights (range 1000 to 1200), unusual for solutes analyzed by GC. The structure of, e.g., the derivative of thyroxine is unambiguous (XV), for T_4Ac we assume the following one (XVI) and the internal standard has the following formula (XVII):

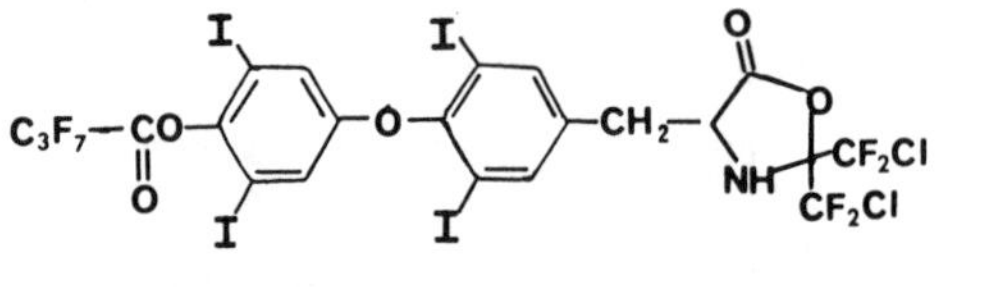

(XV)

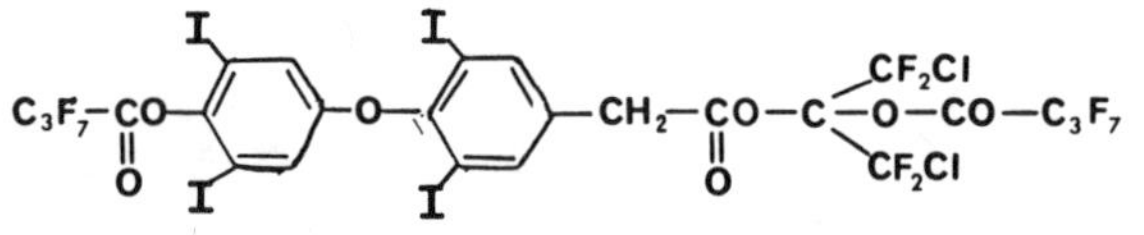

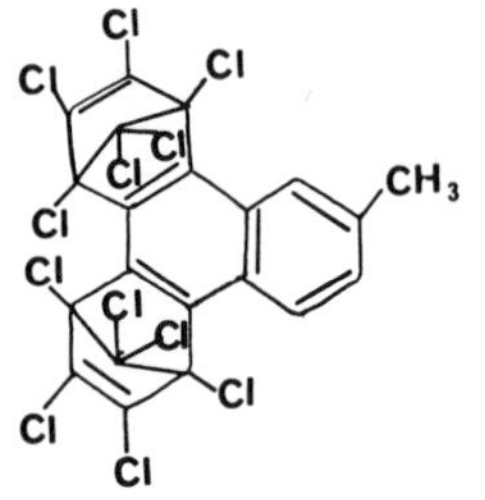

(XVII)

(XVI)

The internal standard undergoes no changes during the chemical treatment and its retention time perfectly suits the required analytic range. It can be easily eluted even from short packed columns, whereas the derivatized thyroidal compounds are completely adsorbed in packed columns. Thus, only the newly introduced inert FSOT columns with immobilized methylsilicone gums are usable for the analysis of the thyroidal compounds. With nitrogen (instead of hydrogen) the results are worse and only hydrogen (or helium) is recommended as carrier gas.

The behavior of the perhalogenated compounds to electron capture detection and the detection limit are currently under study. Amounts down to 10 fmol could be chromatographed without problem with splitless injection or capillary on-column injection. Application of this method for low human serum levels is desirable.

V. CONCLUSIONS

Compared with the esterification procedures, the formation of cyclic derivatives by treatment of amino acids with DCTFA has the following advantages: (1) both characteristic groups, the α-amino and the carboxyl, are blocked by action of a single reagent; (2) the aprotic condensation medium allows one to perform the subsequent acylation step in the same milieu; (3) both reactions proceed very smoothly at room temperature and the acylation is instantaneous; (4) glutamine and asparagine are preserved; (5) histidine, arginine, tryptophan, and cystine can be estimated with a good reproducibility. The only drawback is the requirement for a separate column for determination of the latter amino acids. Substitution of the packed columns by wide-bore capillaries of reasonable length gives the possibility of

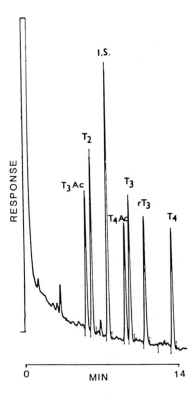

FIGURE 16.   GC analysis of a standard equimolar mixture (20 pmol of each compound injected) of the thyroid substances after treatment with DCTFA and HFBA as described. For the internal standard (I.S.), added in equimolar amount, and the chromatographic conditions see the text. Attenuation: 10^{-11} A.

analysis of the complete spectrum of protein amino acids with a single column in 10 or even 5 min, which is the shortest time reported for amino acid analysis of a complex mixture by GC. Because the derivatives exhibit an enhanced response to the ECD, the iodoamino acids — thyroid hormones and their deamination analogs — might be detectable at serum levels, provided that an efficient cleanup procedure can be found.

However, a basic problem exists: how to obtain the reagent, DCTFA, as at present all the previous distributors have discontinued sale of DCTFA. As the reagent is capable of also converting hydroxyl-containing carboxylic acids and fatty acids to suitable derivatives for GC in tens of seconds (examined at present) its comeback on the market would be certainly profitable. We hope that this conclusion will be helpful in this matter.

ACKNOWLEDGMENTS

Gratitude is expressed to Elsevier Scientific Publishing for permission to publish figures from the *Journal of Chromatography*.

REFERENCES

1. **Hušek, P. and Macek, K.,** Gas chromatography of amino acids, *J. Chromatogr.,* 113, 139, 1975.
2. **MacKenzie, S. L. and Tenaschuk, D.,** Gas-liquid chromatographic assay for asparagine and glutamine, *J. Chromatogr.,* 322, 228, 1985.
3. **Simmons, H. D. and Wiley, D. W.,** I. Fluoroketones, *J. Am. Chem. Soc.,* 82, 2288, 1960.
4. **Engelhardt, K.,** Über aktivierte Ester und Synthese und Gaschromatographie neuer Aminosäurenderivate, Dissertation, TH München, 1963.
5. **Weygand, F.,** Aminosäuren und Peptide, *Z. Anal. Chem.,* 205, 407, 1964.
6. **Weygand, F., Engelhardt, K., Burger, K., and Prox, A.,** unpublished results.
7. **Hušek, P.,** Derivatization of amino acids with 1,3-dichlorotetrafluoroacetone and its use in gas chromatography, *J. Chromatogr.,* 91, 475, 1974.
8. **Hušek, P.,** Gas chromatographic behaviour of amino acid oxazolidinones, response to flame ionization and electron capture detectors, *J. Chromatogr.,* 91, 483, 1974.
9. **Felt, V. and Hušek, P.,** Effect of derivative structure on flame-ionization detector response of amino acid oxazolidinones, *J. Chromatogr.,* 197, 226, 1980.
10. **Hušek, P., Felt, V., and Matucha, M.,** Cyclic amino acid derivatives in gas chromatogrphy, *J. Chromatogr.,* 180, 53, 1979.
11. **Hušek, P.,** Simple method for the determination of histidine, tryptophan, cystine and homocystine by gas chromatography, *J. Chromatogr.,* 172, 468, 1979.
12. **Brooks, J. B., Alley, C. C. and Liddle, J. A.,** Simultaneous esterification of carboxyl and hydroxyl groups with alcohols and heptafluorobutyric anhydride for analysis by gas chromatography, *Anal. Chem.,* 46, 1930, 1974.
13. **Liardon, R., Ott-Kühn, U., and Hušek, P.,** Mass spectra of α-amino acid oxazolidinones, *Biomed. Mass Spectrom.,* 6, 381, 1979.
14. **Hušek, P. and Felt, V.,** Rapid derivatization and gas chromatographic estimation of dicarboxylic amino acids, *J. Chromatogr.,* 152, 363, 1978.
15. **Hušek, P. and Felt, V.,** Simultaneous estimation of dicarboxylic amino acids and their amides by gas chromatography, *J. Chromatogr.,* 152, 546, 1978.
16. **Hušek, P.,** Gas chromatography of cyclic amino acid derivatives — a useful alternative to esterification procedures, *J. Chromatogr.,* 234, 381, 1982.
17. **Hušek, P., Herzogová, G., and Felt, V.,** Contribution to clean-up procedures for serum amino acids, *J. Chromatogr.,* 236, 493, 1982.
18. **Hušek, P., Felt, V., and Matucha, M.,** Single-column gas chromatographic analysis of amino acid oxazolidinones, *J. Chromatogr.,* 252, 217, 1982.
19. **Hušek, P. and Felt, V.,** Possibilities and limitations in the analysis of amino acid oxazolidinones in the femtomole range by gas chromatography with electron-capture detection, *J. Chromatogr.,* 305, 442, 1984.
20. **Hušek, P. and Felt, V.,** A fast derivatization method for the gas chromatographic analysis of thyroid hormonal compounds, *J. Chromatogr.,* 288, 215, 1984.
21. **Hušek, P.,** A highly sensitive electron-capturing standard for application in high-temperature gas chromatography, *J. Chromatogr.,* 288, 200, 1984.

Chapter 6

GAS CHROMATOGRAPHY OF PROTEIN AMINO ACIDS AS THEIR *N*-ISOBUTYLOXYCARBONYL METHYL ESTER DERIVATIVES

Masami Makita and Shigeo Yamamoto

TABLE OF CONTENTS

I. INTRODUCTION

Since gas chromatography (GC) has the general advantages of low cost, versatility, simplicity, speed, and sensitivity of analysis, it has become universally accepted as an analytical method of choice for the determination of mixtures containing organic compounds of biochemical, clinical, chemical, or nutritional interest. However, amino acids are prohibited from directly utilizing this technique owing to their polar and amphoteric character, and consequently many investigators have resorted to the conversion of amino acids to suitable volatile derivatives to accomplish satisfactory GC analysis. Reviews of the volatile derivatives and the techniques involved in obtaining a separation of these derivatives by GC have appeared.[1-3] Continuous efforts have led to the appearance of certain promising derivatives and sophisticated derivatization procedures for precise GC analysis of amino acids; the *N*-acyl alkyl esters including the *N*-trifluoroacetyl (TFA) methyl esters,[4-7] the *N*-TFA *n*-butyl esters,[8-16] the *N*-heptafluorobutyryl (HFB) *n*-propyl esters,[17-20] the *N*-HFB isobutyl esters,[21-31] the N-HFB isoamyl esters,[32,33] the *N*-pentafluoropropionyl (PFP) *iso*- and *n*-propyl esters,[34-36] and *N*-acetyl *n*-propyl esters[37-41] have been studied most extensively, some of which have been practically employed for the successful quantitation of amino acids in biological samples. However, it has been recently pointed out that some of these derivatives possess certain negative aspects when applied to routine laboratory usage.[42]

In 1975, we proposed a new GC approach, in which the volatile derivatives, *N*-isobutyloxycarbonyl (isoBOC) methyl esters, were prepared by reaction with isobutyl chloroformate (isoBCF) in aqueous alkaline media, followed by esterification with diazomethane.[43] The reactions proceed as shown in Figure 1. The derivatives can be easily, rapidly, and quantitatively prepared without the necessity of heating and, needless to say, were stable to moisture, and thus appeared to be useful in practice.[44] Arginine, an exceptional amino acid in our method, was successfully analyzed as the derivative of ornithine after facile treatment with arginase.[44] Recently, the improvement with respect to GC columns, their operating conditions and elimination of the contaminant peaks derived from solvents and reaction vials rendered the method applicable to analysis of submicrogram amounts of all the protein amino acids, including asparagine and glutamine.[45] The method was successfully applied to the determination of amino acids in various biological samples.[46-50]

II. GENERAL PROCEDURE

A. Apparatus and Reagents

A Shimadzu® 4CM gas chromatograph (Shimadzu Seisakushyo, Kyoto, Japan) with a double-column oven with on-column injection ports, two flame ionization detectors (FIDs), a temperature programmer, and two electrometers was employed. Each electrometer was individually equipped with a Shimadzu® R-111 one-pen recorder. A shaker (Iwaki KM Shaker VS Type, KK Iwaki, Tokyo, Japan) set at 300 rpm (up and down) was used for isobutyloxycarbonylation. Pyrex® glass screw top culture tubes (Corning No. 9826) or Wheaton® vials (Wheaton No. 224892 and 224950, Wheaton Scientific, Millville, N.J.) with polytetrafluoroethylene (PTFE)-lined caps were used for isobutyloxycarbonylation, and conical centrifuge tubes (10 mℓ) with screw caps (Iwaki Glass, Tokyo, Japan) for esterification. The reaction vials for isobutyloxycarbonylation were silanized with 5% dimethyldichlorosilane (DMCS) in toluene for 10 min and subsequently washed with methanol.

Standard amino acids and internal standards were obtained from commercial sources, and γ-carboxyglutamic acid monoammonium salt was synthesized according to the procedure of Fernlund et al.[51] After being kept overnight in an evacuated desiccator over phosphorus pentoxide, the standards were weighed to give the stock solutions in 0.1 *M* HCl or in water. Arginase solution was prepared as follows: 10 mg of arginase (40 units/mg, Sigma Chemical

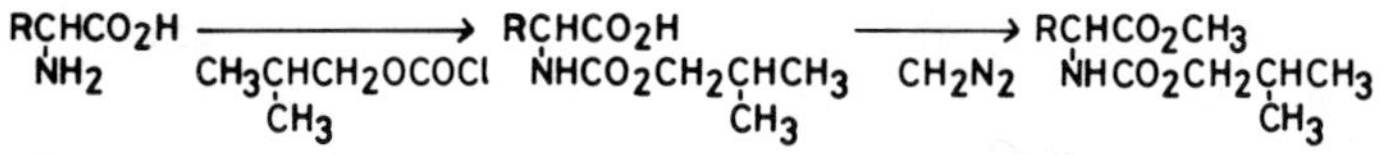

$$RCHCO_2H \xrightarrow{\quad} RCHCO_2H \xrightarrow{\quad} RCHCO_2CH_3$$

RCHCO₂H: NH₂   CH₃CHCH₂OCOCl   ṄHCO₂CH₂CHCH₃   CH₂N₂   ṄHCO₂CH₂CHCH₃
CH₃ ... CH₃ ... CH₃

FIGURE 1.   Derivatization of amino acids to the *N*-isobutoxycarbonyl (iso-BOC) methyl esters.

Co., St. Louis, Mo.) were activated in 0.4 mℓ of 1.25 *M* ammonium acetate and 0.1 mℓ of 0.05 *M* manganese (II) sulfate at 37°C for 4 hr. After centrifugation for 1 min at 2200 × *g*, the supernatant was separated, to which 0.5 mℓ of water was added. This solution could be stored frozen without loss of activity for at least 15 days. Isobutylchloroformate (IsoBCF) stabilized with calcium carbonate and ethyl chloroformate (ECF) were obtained from Tokyo Kasei Kogyo (Tokyo, Japan) and refrigerated when not in use. *N*-Methyl-*N*-nitroso-*p*-toluenesulfonamide and diethylene glycol monoethyl ether for the generation of diazomethane were obtained from Wako Pure Chemical (Osaka, Japan). Diethyl ether (100 mℓ) was treated three times with 10 mℓ of acidic iron (II) sulfate solution [$FeSO_4 \cdot 7H_2O$ (120 g) + 95% H_2SO_4 (12 mℓ) + H_2O (220 mℓ)] in order to remove peroxide, and distilled in an all-glass apparatus after being washed with water and dried over anhydrous sodium sulfate. Unless otherwise noted, the water purified as follows was used: deionized water was distilled in an all-glass system after addition of several pellets of sodium hydroxide to remove the acidic contaminants. Sodium sulfate (anhydrous) and sodium chloride were washed with methanol, then with purified diethyl ether and dried at 100°C. All other chemicals and solvents were the purest grades available from standard commercial sources.

B. Preparation of Column Packings

Poly-I-110, Poly-A-101A, FFAP, OV-17 (Applied Science Labs., State College, Pa.) and SP-1000 (Supelco Inc., Bellafonte, Pa.) were used as liquid phases. Uniport® HP, 100/120 mesh (Gasukuro Kogyo, Tokyo, Japan) and Gas-Chrom® P, 100/120 mesh (Applied Science Labs) were used as solid supports. Prior to coating, Uniport® HP was purified and resilanized as follows: the solid support was slowly added to concentrated HCl placed in a separatory funnel and complete contact with the liquid was ensured by gentle swirling. The grey and black particles, which were precipitated, and the acid were separated. After this procedure was repeated three more times, the solid support was again floated on deionized water and a similar procedure to that described above was carried out until neutral to remove further grey and black particles as well as the remaining acid. The solid support, which was transferred into a flask with methanol and was slurried with the addition of methanol, was then washed four times with the same solvent with a slight agitation of the flask. At each washing with methanol the fine particles were removed by decantation. The solid support soaked in methanol was collected on a glass filter (G 1). After filtration and drying at 100°C, the solid support was silanized with 5% DMCS in toluene.[52] Prior to coating, Gas-Chrom® P was only resilanized in the same way as described above.

The column packings used were prepared by the filtration technique[52] using 1-butanol/chloroform (1:1, v/v) as coating solvents. In the case of a mixed liquid phase, 1.605% Poly-I-110/Poly-A-101A/FFAP (1200:300:105, w/w/w), the solution obtained after dissolving the liquid phases in the coating solvents by vigorous shaking for 1 hr was filtered to remove insoluble materials. The 1.605% Poly-I-110/Poly-A-101A/FFAP and 1.0% Poly-A-101A/FFAP (1:1, w/w) were coated on the purified and resilanized Uniport® HP, and the other liquid phases either on the Uniport® HP or on the resilanized Gas-Chrom® P. The interior of the glass columns and the quartz wool plugs, placed in each end to hold the column packing in place, were silanized with 5% DMCS in toluene. After silanization, they were contacted with methanol for 20 min and successively rinsed with the same solvent to exclude HCl produced during treatment.

The column packings were poured into the clean and dry analytical columns with gentle tapping under suction by an aspirator and the packed columns were conditioned at 280°C (after 2°C/min linear increase from room temperature) for at least 20 hr with a nitrogen flow rate of 30 mℓ/min. Nitrogen (>99.99%) was passed through a tube (20 cm × 3 cm I.D.) containing molecular sieve 5A. Other GC conditions are given in each figure.

C. Preparation of Derivatives

Aliquots of the aqueous standard amino acid mixture or a sample solution (containing up to a total of 2 mg of amino acids, 0.5 to 100 µg of each amino acid) and the internal standard solution were pipetted into a vial (a fixed amount of an appropriate internal standard is added to a sample solution at an appropriate stage of sample pretreatment). To this solution, 0.5 mℓ of 10% sodium carbonate solution was added and the total reaction volume was made up to 2 mℓ with water if necessary. Immediately after 20 µℓ of isoBCF were added, the mixture was shaken with a shaker for 10 min at room temperature. For addition of isoBCF, 10- to 50-µℓ syringes (Hamilton, Reno, Nev.) were employed. The reaction mixture was extracted twice with 3 mℓ of diethyl ether in order to remove the excess reagent and the ether extracts were discarded. This procedure also serves to exclude amines and phenols, both of which often coexist with amino acids in biological materials, as they are derivatized to the corresponding *N*- or *O*-isoBOC derivatives which are soluble in organic solvents under the same conditions as described above.[53-55] After saturation with sodium chloride, the aqueous layer was acidified to pH 1 to 2 with 10% orthophosphoric acid under continuous mixing and then extracted five times with 2 mℓ of diethyl ether with vigorous shaking by hand for 1 min. The ether layers were transferred into another vial by means of a Pasteur capillary pipet (Corning No. 7095B). Saturation with sodium chloride and multiple extraction improved extraction efficiency for the isoBOC amino acids of threonine, serine, hydroxyproline, asparagine, and glutamine, particularly when small amounts were derivatized. After being dried over a few grains of anhydrous sodium sulfate, the combined ether extract was transferred into an esterification vial and esterified without addition of methanol by bubbling diazomethane, generated according to the microscale procedure of Schlenk and Gellerman,[56] through this solution until a yellow tinge persisted. After standing for 5 min at room temperature, the solvent was evaporated almost to dryness at 40 to 45°C without use of a stream of nitrogen. The residue on the walls of the vial was washed down with a small volume of diethyl ether and the solvent was evaporated again to complete dryness. The residue was reconstituted with 20 to 100 µℓ of ethyl acetate and 1 to 5 µℓ of the solution were injected onto the gas chromatograph. The peak height ratios relative to the internal standard were calculated.

Since it has proved impossible to obtain the volatile derivative of arginine by direct use of the derivatization method described above, some modifications to the procedure originally devised by Coulter and Hann[37,38] were introduced for the simple and rapid conversion of arginine into ornithine with arginase. The standard amino acid mixture or a sample solution was placed in a 5-mℓ centrifugation tube, adjusted to pH 9.4 to 9.6 with 5% sodium carbonate solution (after addition of the internal standard solution to the standard amino acid mixture) and made up to about 1.5 mℓ with water if necessary. To this solution was added 60 µℓ of the arginase solution (ca. 24 units per vial). After incubation for 5 min at 37°C with occasional shaking, 0.5 mℓ of a 10% sodium carbonate solution was added and the precipitates formed were centrifuged off at 2200 × *g* for 3 min. The supernatant was transferred into a reaction vial and was subjected to the derivatization procedure described above.

D. Comments on the General Procedure

1.   When small amounts of amino acids (<2 µg of each) were derivatized, losses of some amino acids, notably of histidine, due to adsorption to surfaces of the reaction vials

were observed. This problem was avoided by silanizing the vials used in the isobutyloxycarbonylation.

2. Before use, the glassware for derivative preparation (reaction vial, Pasteur capillary pipet, and measuring pipet) should be washed with distilled water then rinsed with methanol and then with acetone. Cleaning glassware with synthetic detergents should be avoided because an adverse effect on tyrosine derivatization was sometimes observed, although we have no explanation for this event.

3. IsoBCF (bp, 128°C) stabilized with calcium carbonate obtained from Tokyo Kasei Kogyo could be used without further purification. However, if other products must be used, the purity should be checked because we experienced that some other products were too low in purity to use.

4. Peroxide-free diethyl ether should be employed to obviate decomposition of methionine. Greater care should be taken for this solvent when small amounts of methionine are to be analyzed.

5. Blanks throughout the derivatization procedure should be run to confirm whether contaminant peaks are present. Gaskets of reaction vials and water mainly caused the contaminant peaks. Synthetic resin gaskets other than PTFE-lined should not be used.

6. The ether layers should be collected, taking care to avoid aqueous droplets. It was not necessary to completely draw the ether layer in each extraction.

7. It should be noted that diazomethane is explosive and toxic. To minimize leakage and glassware breakage, a small-scale generation system of diazomethane is advisable. Recently, a convenient apparatus has been presented.[57]

8. When the Uniport® HP support material was used as received, quantification of asparagine, glutamine, and cystine was often insufficient, indicating breakdown of these derivatives on the column. This was the case with Gas-Chrom® P. Purification of the solid support provided increased responses for these amino acids and made it possible to quantify them.

9. The total reaction volume of the isobutyloxycarbonylation could be reduced to below 1 mℓ while the concentration of sodium carbonate was maintained at 2.5% (w/v). From experiments with the standard mixture containing 100 μg of each protein amino acid, it was found that the amount of isoBCF equivalent to about 1% (v/v) of the total reaction volume was enough to quantitatively and reproducibly accomplish the reaction.

10. For analysis of free amino acids in samples containing proteins, we recommend the use of perchloric acid for prior deproteinization.

III. SEPARATION AND DETERMINATION OF 22 PROTEIN AMINO ACIDS

A. Nature of Derivatives

A series of *N*-alkyloxycarbonyl methyl esters of the protein amino acids was prepared by the use of various alkyl chloroformates (ethyl, *n*-butyl, isobutyl, *n*-amyl, *n*-hexyl, and *n*-octyl) according to the derivatization method described above, and they were evaluated in terms of their volatility and quantification. Methyl chloroformate was discarded because of its instability under the reaction conditions investigated. As expected, the *N*-ethyloxycarbonyl (EOC) methyl esters were found to be the most volatile, but attempts to use this derivative for the complete GC analysis of all the protein amino acids were found to be unsuccessful because of incomplete extraction of the *N*-EOC amino acids of threonine, serine, hydroxyproline, asparagine, and glutamine, particularly when working at low levels. The other derivatives, except the *N*-isoBOC methyl esters, were found to be insufficiently volatile for use in the complete analysis of the protein amino acids. As a consequence of these experiments, the *N*-isoBOC methyl esters were selected for the determination of the 22 protein amino acids including asparagine and glutamine. However, the other derivatives are complementary, in other words some nonprotein amino acids could be separated as these derivatives, and they may be useful as an alternative for gas chromatographic/mass spectrometric (GC/MS) identification. Recently, as *n*-propyl chloroformate has become available from Tokyo Kasei Kogyo, the *N*-*n*-propyloxycarbonyl methyl esters were prepared and analyzed by GC to compare with the *N*-isoBOC methyl esters. The former showed a similar behavior in volatility and separation to the latter, but, like *N*-EOC methyl esters, there were difficulties in quantitative determination of asparagine and glutamine.

The volatility of the *N*-isoBOC methyl esters are modest compared with the *N*-perfluoroacyl alkyl esters, and this was particularly helpful in preventing losses of the more volatile derivatives during evaporation of solvent. The salient feature of the proposed derivatives is their stability towards moisture, and therefore no special precautions to exclude moisture are necessary in their handling and storage. If it was necessary to store the samples either as in solid state or in solution in ethyl acetate, they could be kept in a refrigerator (4°C) for at least 1 week without any destruction of the derivatives. However, degradation, probably oxidative, of the derivatives of methionine, cysteine, and tryptophan clearly occurred with more prolonged storage.

The structures of the *N*-isoBOC methyl esters of the protein amino acids prepared by the derivatization method were elucidated by GC/MS. Molecular ion peaks which are consistent with the structures postulated were observed for all protein amino acids. The GC/MS study suggested that: (1) all carboxyl groups are esterified; (2) all amino, imino, imidazole, phenolic hydroxyl, and sulfhydryl groups are substituted with isoBOC groups, whereas alcoholic hydroxyl groups and indole ring nitrogen as well as amide groups are not. In addition to the GC/MS study, the structures of the derivatives of some representative amino acids including threonine, cysteine, tyrosine, and tryptophan were verified by elemental analyses.[44]

B. Separation of the *N*-isoBOC Methyl Esters

The chromatographic and instrumental conditions for the separation of the *N*-isoBOC methyl esters of the common protein amino acids were investigated in detail. Among the liquid phases used for screening were OV-1, OV-17, OV-25, neopentyl glycol succinate, cyclohexane dimethanol succinate, Versamide 930, FFAP, Poly-I-110, Poly-A-101A, and Poly-A-103. The polar liquid phases generally afforded better separation and better quantification than did the nonpolar or slightly polar liquid phases such as OV-1 and OV-17, although the latter gave a higher response to cystine (Figure 2). It was found difficult to achieve a complete separation by using a single liquid phase, and also it was found necessary to use polar liquid phases with good stability at high temperatures. Several mixed liquid phases were therefore investigated on the basis of the results obtained with the liquid phases described above. Consequently, it was found that a 1.605% Poly-I-110/Poly-A-101A/FFAP column (column A) provided the desired separation of all the protein amino acids except isoleucine and leucine, which were co-eluted, and on the other hand a 1.0% Poly-A-101A/FFAP column (column B) gave a good enough separation of this pair even when operated under the same thermal conditions as employed for the former column (Figure 3).

Our method necessitated a separate analysis if quantitation of ornithine was required at the same time as arginine was being assayed. In addition, a problem arose with the quantitative determinations of methionine, cysteine, and tryptophan when test mixtures were subjected to the arginase treatment. Their GC responses, notably of cysteine, were reduced by oxidative degradation which may be caused by manganese (II) during incubation with arginase. Therefore, in view of the problems mentioned above, we decided to operate both columns simultaneously (not differentially) and to assay isoleucine and leucine as well as arginine, which was already converted into ornithine, on column B. In practice, in a sample containing both arginine and ornithine, the amount of arginine was determined by subtracting

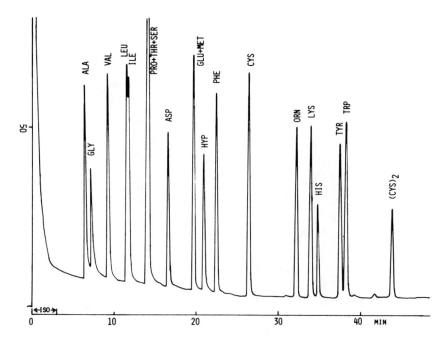

FIGURE 2. Separation of 20 amino acids as *N*-isoBOC methyl esters on a 1.0% OV-17 column. Conditions: column, 2 m × 3 mm I.D.; solid support, 100/120 mesh Gas-Chrom® P; temperature program, isothermal at 100°C for 3 min, then programmed at 4°/min to 270°C; nitrogen flow rate, 40 mℓ/min.

the amount of ornithine determined without arginase treatment using column A from that determined with arginase treatment using column B, and then by converting the remaining ornithine into arginine. In this method, two derivatizations and two injections per sample were necessary for complete analysis of the protein amino acids. The simultaneous operation of two columns, however, saved the time needed for GC analysis.

C. Quantitative Aspects

Pure reference derivatives of nine representative amino acids (valine, aspartic acid, threonine, phenylalanine, cysteine, ornithine, lysine, tyrosine, and tryptophan) were synthesized in order to assess the derivatization yields.[44] The results of elemental analyses of these pure references were in agreement with the calculated values. The derivatization yields of the nine amino acids were above 95% in each instance when the standard mixture containing the 22 protein amino acids plus ornithine (100 μg of each) was derivatized under the conditions described above.

The conversion yield of arginine into ornithine by arginase was almost quantitative in the range 0.5 to 100 μg of arginine, and the presence of the other protein amino acids did not interfere with the enzymatic reaction.[44,45]

Calibration graphs for all protein amino acids proved to be linear and reproducible in the range 0.5 to 100 μg.[44,45] Histidine showed a good peak, and required no extra derivatization step, unlike the *N*-perfluoroacyl alkyl esters.[26,34,58]

IV. APPLICATIONS

A. Protein Hydrolysates[46]

Bovine insulin, α-chymotrypsin, albumin, and milk casein were used to explore the applicability of the method for determining amino acid compositions. After being kept

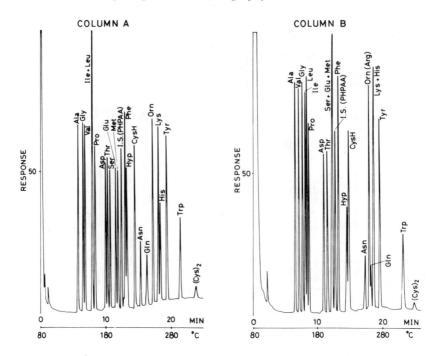

FIGURE 3.   Separation of 22 amino acids as *N*-isoBOC ethyl esters on a dual set of columns. Column A: 1.605% Poly-I-110/Poly-A-101A/FFAP (1200:300:105, w/w/w) on 100/120 mesh Uniport® HP, 1 m × 3 mm I.D. Column B: 1.0% Poly-A-101A/FFAP (1:1, w/w) on 100/120 mesh Uniport® HP, 1 m × 3 mm I.D. Conditions: injection and detector temperatures, 280°C; nitrogen flow rate, 35 mℓ/min; hydrogen flow rate, 50 mℓ/min; air flow rate, 800 mℓ/min; temperature program, linear rise at 10°/min from 80 to 280°C, then held for 5 min. Arginine was separated as the ornithine derivative. Internal standard (I.S.), *p*-hydroxyphenylacetic acid. (From Makita, M., Yamamoto, S., and Kiyama, S., *J. Chromatogr.*, 237, 279, 1982. With permission.)

overnight in an evacuated desiccator over phosphorus pentoxide, 10 mg of each protein sample were weighed into a Pyrex® glass tube (13 × 1.5 cm), and 10 mℓ of 6 *N* HCl were added to the sample. The solution was degassed under vacuum with cooling by a freezing mixture (ice + NaCl), and the tube was flushed with nitrogen before closure. Hydrolysis was conducted for 24 hr at 115°C, but for insulin for 20 hr at 106°C in 6 *N* HCl containing 2% (v/v) phenol.[18] The hydrolyzate solution was filtered through a glass filter and the tube was washed with two 5-mℓ volumes of water. The combined filtrate was taken to dryness with a rotary evaporator at 50°C under reduced pressure. After addition of 0.5 mℓ of the internal standard solution *p*-hydroxyphenylacetic acid, 500 μg/mℓ, to the residue, the whole was transferred with water into a 10-mℓ volumetric flask and brought to volume with water. Two 250-μℓ portions of this solution (corresponding to 250 μg of protein) were placed into the reaction vials, and one was directly derivatized and the other was derivatized after arginase treatment. Gas chromatograms obtained from the insulin hydrolysate are shown in Figure 4. Table 1 shows the comparison of the amino acid molar ratios obtained from the present method with those obtained by the automated amino acid analyzer, indicating that there is a fairly good agreement between the two types of analyses. Further, this table illustrates the accuracy of GC analysis. The data obtained from the other protein hydrolysates were also in good agreement with those obtained by the amino acid analyzer.

B. Serum Amino Acids[50]

Our derivatization method has the advantage that amino acids in aqueous solution can be

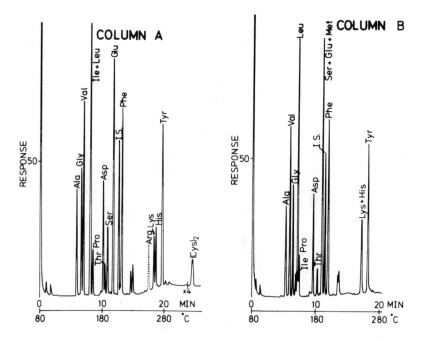

FIGURE 4.   Gas chromatograms of the *N*-isoBOC methyl esters of amino acids in an insulin hydrolysate. Conditions are given in Figure 3. I.S.: *p*-hydroxyphenylacetic acid.

Table 1
MOLAR RATIO (MR) OBTAINED
IN THE AMINO ACID ANALYSES
OF AN INSULIN HYDROLYSATE

Amino acid	GC[a] MR	Amino acid analyzer[b] MR	Theory MR
Ala	2.90	3.04	3
Val	4.18	4.23	5
Gly	3.98	4.26	4
Ile	0.47	0.45	1
Leu	6.16	6.21	6
Pro	1.00	1.02	1
Asp	3.26	3.09	3
Glu	6.48	7.19	7
Ser	3.16	2.81	3
Met	—	—	—
Phe	2.94	2.94	3
His	1.91	2.04	2
Lys	1.17	1.00	1
Tyr	3.92	4.06	4
Cys-Cys	2.65	2.23	3
Arg	0.97[c]	1.02	1

[a]   Each value represents an average of two determinations.

[b]   Hitachi, KLA-5 Amino Acid Analyzer.

[c]   Arginine was converted to ornithine with arginase and analyzed as the derivative of ornithine.

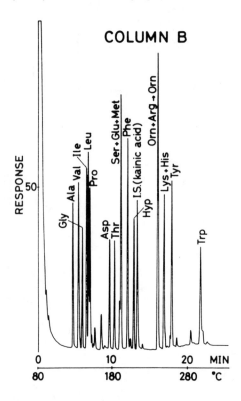

FIGURE 5.   Separation of amino acids as *N*-EOC methyl esters on column B. Column B and conditions are given in Figure 3. (From Yamamoto, S., Kiyama, S., Watanabe, Y., and Makita, M., *J. Chromatogr.*, 233, 39, 1982. With permission.)

converted to their *N*-isoBOC amino acids, which are selectively extracted from the acidified reaction mixture into diethyl ether, without any step to exclude water (moisture) being necessary prior to derivatization. In an attempt to analyze serum amino acids, we aimed at introducing a serum treatment involving neither ion exchange column chromatography (IEC) nor subsequent evaporation. For this purpose, the use of perchloric acid proved to be convenient for the effective and reliable removal of serum proteins coupled with prior extraction of lipids with chloroform from the acidified serum. However, the overall recovery for tyrosine was below 70%, although the other amino acids, including asparagine and glutamine, showed good recovery values. A similar phenomenon was also observed even when serum was cleaned up by picrate precipitation of protein followed by cation-exchange column chromatography.[47] This problem was solved by introduction of the *N*-EOC methyl esters for the determination of leucine, isoleucine, and arginine along with tyrosine, because it was found that the recovery of tyrosine was sufficiently improved when analyzed as its *N*-EOC methyl ester and that the *N*-EOC methyl esters of these four amino acids provided good separation on column B, as shown in Figure 5, and good calibration linearity for each.

Sample preparation was carried out as follows. To 200 μℓ of serum in a glass tube (6.5 cm × 9 mm I.D.) were successively added 100 μℓ of the internal standard solution (kainic acid; 2-carboxy-4-isopropenyl-3-pyrrolidineacetic acid, 50 μℓ/mℓ), 100 μℓ of water and 80 μℓ of 10% orthophosphoric acid. The solution was gently mixed for 10 sec with an equal volume of chloroform with a Vortex®-type mixer and centrifuged at 2200 × *g* for 5 min. The upper layer was transferred to another tube, taking care not to collect any chloroform droplets, and 80 μℓ of 12% perchloric acid were added to the aqueous layer with gentle

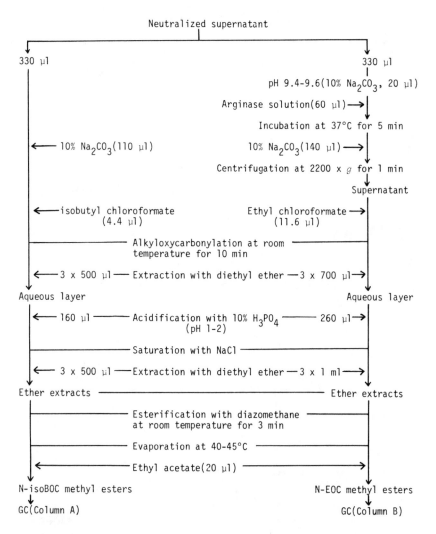

FIGURE 6. Schematic flow diagram of the procedure. (From Yamamoto, S., Kiyama, S., Watanabe, Y., and Makita, M., *J. Chromatogr.*, 233, 39, 1982. With permisson.)

swirling to precipitate proteins. Vigorous mixing should be avoided at this stage. After centrifugation for 1 min at 2500 × *g*, the supernatant was carefully transferred to a small glass tube and neutralized with 10% sodium carbonate (about 120 μℓ). Each 330-μℓ portion of the neutralized supernatant was placed in two 2-mℓ vials (Wheaton No. 224950), and they were treated according to the flow diagram shown in Figure 6. Extraction with diethyl ether was achieved with a Vortex®-type mixer for 30 sec. The resulting derivatives, the *N*-isoBOC methyl esters and the *N*-EOC methyl esters, were analyzed simultaneously on columns A and B, respectively.

Gas chromatograms obtained from a serum sample are shown in Figure 7. Only a few and small extraneous peaks are detected on the chromatograms, thus indicating that the procedure provides adequate purification and separation of amino acids in sera. As can be seen, the amides emerged as symmetrical and well-separated peaks and could be determined separately from the respective acids and also from all other amino acids. Quantitative aspects, i.e., recovery rates of amino acids added to serum and reproducibility of analysis, were sufficient to perform quantitative analyses of serum amino acids.

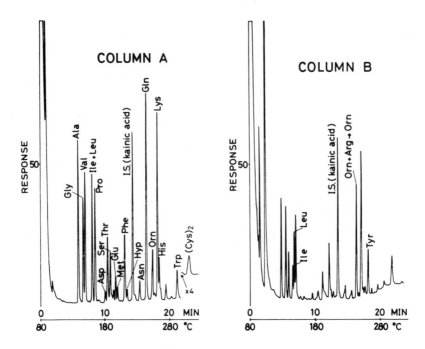

FIGURE 7.   Representative gas chromatograms of the *N*-isoBOC methyl esters (Column A) and the *N*-EOC methyl esters (Column B) of amino acids in a serum sample. Prior to preparation of the *N*-EOC methyl esters, a serum sample was treated with arginase, and only the peaks labeled by their names were determined. The nitrogen flow rates for Columns A and B are 35 and 25 mℓ/min, respectively. Other conditions are the same as in Figure 3. (From Yamamoto, S., Kiyamo, S., Watamabe, Y., and Makita, M., *J. Chromatogr.*, 233, 39, 1982. With permission.)

C. Hydroxyproline in Urine Hydrolyzates[48]

It has been found that the urinary excretion of hydroxyproline (Hyp) can be used as an indicator of various diseases associated with collagen metabolism.[59]

We have developed a simple and specific GC method for the determination of Hyp in urine hydrolysates. A urine specimen was hydrolyzed at 100°C for 22 hr in a closed tube with 6 *N* HCl under nitrogen pressure, and after removal of the HCl Hyp was converted to its *N*-isoBOC methyl ester without any cleanup procedure. Hyp was clearly separated from other urinary constituents on a 0.6% FFAP column (Figure 8) and analyzed with high accuracy.

D. Analysis of Urinary Free γ-Carboxyglutamic Acid[49]

γ-Carboxyglutamic acid (Gla) has been found to be present in several calcium-binding proteins and to play important roles in blood coagulation and calcification. It has been demonstrated that urinary Gla is a product of degradation of the Gla-containing proteins.[60,61] These results suggested that the urinary Gla level is a useful indicator to assess the diseases associated with blood coagulation and bone metabolism.

A reliable GC method has been developed for the determination of urinary free Gla, which takes advantage of the mild derivatization conditions of our method; Gla is highly acid labile and is converted to glutamic acid by contact with acid, and this suggests that the methods which require esterification with HCl and alcohols under drastic conditions cannot be applied to estimation of this amino acid.

Sample pretreatment was carried out as follows: 10 mℓ of the supernatant obtained from a 24-hr urine collection was placed in a 50-mℓ centrifuge tube with a ground glass stopper

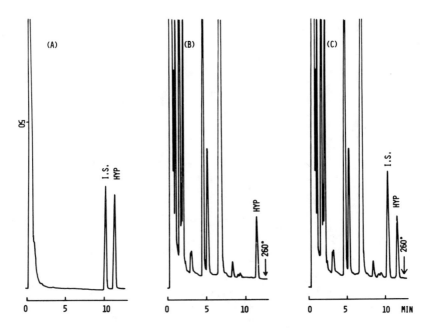

FIGURE 8.   Gas chromatograms obtained from (A) standard, (B) urine hydrolysate, and (C) urine hydrolysate plus internal standard (I.S.), kainic acid. Conditions: column, 0.6% FFAP on 100/120 mesh silanized Gas-Chrom® P (2 m × 3 mm I.D.); temperature program, isothermal at 170°C for 1 min, then programmed at 4°C/min to 215°C (soon after elution of Hyp, the column temperature was raised to 260°C by manual operation and held for 5 min); nitrogen flow rate, 40 mℓ/min. (From Makita, M., Yamamoto, S., and Tsudaka, Y., *Clin. Chim. Acta,* 88, 305, 1978. With permission.)

and after adjusting to pH 1 to 2 with 20% HCl was extracted five times with 10 mℓ of diethyl ether by shaking for 5 min to remove acidic components. The aqueous layer was then adjusted to pH 8 to 9 with 10% sodium hydroxide. After removal of the resulting precipitates by centrifugation at 1500 × *g* for 3 min, the supernatant was adjusted to pH 6.5 with 20% HCl and this solution was transferred to a 50-mℓ volumetric flask, then brought to volume with water. A 5-mℓ volume of this solution (corresponding to 1 mℓ of 24 hr urine) was loaded onto a column of Amberlite® CG-120 (7 mm × 5 cm, H⁺ form) and then eluted with water. Gla was not retained due to its highly negative charge. The portion of the eluate from the beginning of the ion exchange cleanup operation was collected, and to this eluate was added 0.4 mℓ of internal standard solution (*N*-isoBOC tranexamic acid, 100 μg/mℓ). The eluate was evaporated to dryness at 60°C at reduced pressure with a rotary evaporator. The residue was transferred to a reaction vial with 2.5 mℓ of 2.5% sodium carbonate solution and was derivatized to the *N*-isoBOC trimethyl ester.

The derivative of Gla showed a single and symmetrical peak and was clearly separated from other urinary constituents on a 1.5% OV-17/0.2% SP-1000 column (Figure 9). No degradation of Gla to glutamic acid was observed throughout the overall procedure. The relative standard deviations of peak height ratios were 1.05 to 3.66% when Gla was derivatized at the 1- to 100-μg levels. The recovery from urine samples fortified with 5 to 100 μg of Gla were 91.6 to 98.3%. The urinary excretion of Gla in normal subjects determined by the present method ranged from 13.1 to 68.0 μmol/24 hr, which were in agreement with those determined by an automated amino acid analyzer.[60,61]

E. Analysis of Nonprotein Amino Acids[62]

The derivatization method was also applicable to the analysis of nonprotein amino acids such as shown in Figure 10. Among nonprotein amino acids investigated, taurine, citrulline,

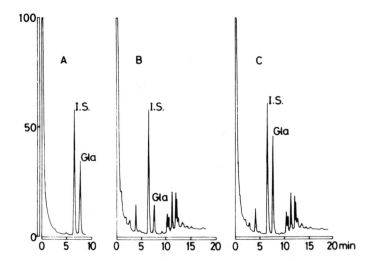

FIGURE 9.   Gas chromatograms obtained from (A) standard Gla, 30 μg; internal standard (I.S.), 40 μg of *N*-isoBOC tranexamic acid; (B) urine (1 mℓ of 24-hr collection), and (C) urine fortified with 30 μg of Gla. Conditions: column, 1.5% OV-17/0.2% SP-1000 on 100/120 mesh Uniport® HP (0.5 m × 3 mm I.D.); temperature program, programmed from 150°C at 6°/min (soon after elution of Gla, the temperature was raised to 275°C by manual operation and held for 6 min); nitrogen flow rate, 40 mℓ/min. (From Matsu-ura, S., Yamamoto, S., and Makita, M., *Anal. Biochem.*, 114, 371, 1981. With permission.)

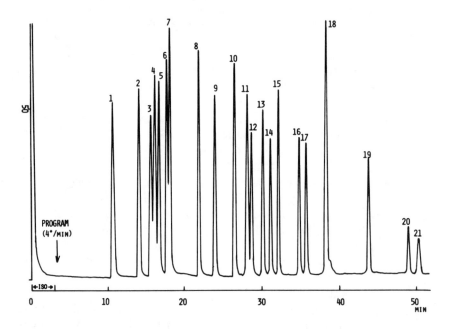

FIGURE 10.   Gas chromatogram of some nonprotein amino acids as *N*-isoBOC methyl esters. Conditions: column, 0.5% FFAP on 100/120 mesh silanized Gas-Chrom® P (2 m × 3 mm I.D.); temperature program, isothermal at 80°C for 3 min. Then programmed at 4°/min to 275°C; nitrogen flow rate, 40 mℓ/min. Peaks: 1, sarcosine; 2, α-amino-*n*-butyric acid; 3, *allo*isoleucine; 4, norvaline; 5, β-aminoisobutyric acid; 6, β-alanine; 7, norleucine; 8, γ-aminobutyric acid; 9, *S*-methylcysteine; 10, ε-aminocaproic acid; 11, ethionine; 12, homoserine; 13, α-aminoadipic acid; 14, δ-aminolevulinic acid; 15, kainic acid; 16, *S*-carboxymethylcysteine; 17, homocysteine; 18, 2,4-diaminobutyric acid; 19, methionine sulfone; 20, lanthionine; 21, δ-hydroxylysine. (From Makita, M., Yamamoto, S., Sakai, K., and Shiraishi, M., *J. Chromatogr.*, 124, 92, 1976. With permission.)

and cysteic acid gave no peaks. The quantitative aspect was satisfactory for each nonprotein amino acid investigated.

V. SUMMARY

Rapid and accurate amino acid analysis by GC has been clearly demonstrated with the *N*-isoBOC methyl ester derivatives of the 22 protein amino acids including asparagine and glutamine. The derivatization reaction consists of two steps: alkyloxycarbonylation with isobutyl chloroformate (and ethyl chloroformate) in an aqueous alkaline medium and subsequent esterificiation with diazomethane. These chemical reactions proceed under mild conditions without the necessity of heating. Further, the resulting derivatives are so stable to moisture that no special handling and precaution are necessary, and these features are of particular importance when the method is applied to routine laboratory usage. Only 20 min is needed for the derivatization and several samples can be completed simultaneously. The GC analysis of the 22 protein amino acids can be performed in 25 min by the use of a dual set of columns with the same thermal conditions.

To determine arginine, arginase treatment is introduced to convert arginine to ornithine prior to derivatization, because arginine gave no volatile derivative under the derivatization conditions that were suitable for the other amino acids. In this method, the complete separation of the 22 protein amino acids requires a two-column system and the peak height for cystine is relatively small. In order to overcome these shortcomings, we are at present studying the possibility of improving the columns.

REFERENCES

1. **Weinstein, B.,** Separation and determination of amino acids and peptides by gas-liquid chromatography, *Methods Biochem. Anal.,* 14, 203, 1966.
2. **Blau, K.,** Analysis of amino acids by gas chromatography, in *Biomedical Applications of Gas Chromatography,* Vol. 2, Szymanski, H. A., Ed., Plenum Press, New York, 1968, 1.
3. **Hušek, P. and Macek, K.,** Gas chromatography of amino acids, *J. Chromatogr.,* 113, 139, 1975.
4. **Cruickshank, P. A. and Sheehan, J. C.,** Gas chromatographic analysis of amino acids as *N*-trifluoroacetyl amino acid methyl esters, *Anal. Chem.,* 36, 1191, 1964.
5. **Darbre, A. and Islam, A.,** Gas-liquid chromatography of trifluoroacetylated amino methyl esters, *Biochem. J.,* 106, 923, 1968.
6. **Islam, A. and Darbre, A.,** Gas-liquid chromatography of trifluoroacetylated amino acid methyl esters. Development of a mixed stationary phase for their separation, *J. Chromatogr.,* 71, 223, 1972.
7. **Cliffe, A. J., Berridge, N. J., and Westgarth, D. R.,** Determination of some amino acids by gas chromatography of derivatives, *J. Chromatogr.,* 78, 333, 1973.
8. **Zomzely, C., Marco, G., and Emery, E.,** Gas chromatography of the *n*-butyl-*N*-trifluoroacetyl derivatives of amino acids, *Anal. Chem.,* 34, 1414, 1962.
9. **Lamkin, W. M. and Gehrke, C. W.,** Quantitative gas chromatography of amino acids. Preparation of *n*-butyl *N*-trifluoroacetyl esters, *Anal. Chem.,* 37, 383, 1965.
10. **Gehrke, C. W., Zumwalt, R. W., and Wall, L. L.,** Gas-liquid chromatography of protein amino acids. Separation factors, *J. Chromatogr.,* 37, 398, 1968.
11. **Roach, D. and Gehrke, C. W.,** The gas-liquid chromatography of amino acids, *J. Chromatogr.,* 43, 303, 1969.
12. **Zumwalt, R. W., Roach, D., and Gehrke, C. W.,** Gas-liquid chromatography of amino acids in biological substances, *J. Chromatogr.,* 53, 171, 1970.
13. **Gehrke, C. W., Zumwalt, R. W., and Kuo, K.,** Quantitative amino acid analysis by gas-liquid chromatography, *J. Agric. Food Chem.,* 19, 605, 1971.
14. **Pellizzari, E. D., Brown, J. H., Talbot, P., Farmer, R. W., and Fabre, L. F., Jr.,** An evaluation of the gas chromatographic analysis of plasma amino acids, *J. Chromatogr.,* 55, 281, 1971.

15. **Tucker, H. N. and Molinary, S. V.,** Gas chromatography in diagnostic biochemistry of abnormal valine metabolism, *Clin. Chem.,* 19, 1040, 1973.

16. **Kaiser, F. E., Gehrke, C. W., Zumwalt, R. W., and Kuo, K. C.,** Amino acid analysis: hydrolysis, ion-exchange clean-up, derivatization, and quantitation by gas-liquid chromatography, *J. Chromatogr.,* 94, 113, 1974.

17. **Moss, C. W., Lambert, M. A., and Diaz, F. J.,** Gas-liquid chromatography of twenty protein amino acids on a single column, *J. Chromatogr.,* 60, 134, 1971.

18. **Jönsson, J., Eyem, J., and Sjöquist, J.,** Quantitative gas chromatographic analysis of amino acids on a short glass capillary column, *Anal. Biochem.,* 51, 204, 1973.

19. **Kirkman, M. A.,** Comparative determination of protein amino acids in plant materials by automated cation exchange and gas-liquid chromatography of the amino acid N-heptafluorobutyryl, n-propyl esters, *J. Chromatogr.,* 97, 175, 1974.

20. **March, J. F.,** Modified technique for the quantitative analysis of amino acids by gas chromatography using heptafluorobutyric n-propyl derivatives, *Anal. Biochem.,* 69, 420, 1975.

21. **MacKenzie, S. L. and Tenaschuk, D.,** Gas-liquid chromatography of N-heptafluorobutyryl isobutyl esters of amino acids, *J. Chromatogr.,* 97, 19, 1974.

22. **MacKenzie, S. L. and Tenaschuk, D.,** Rapid formation of amino acid isobutyl esters for gas chromatography, *J. Chromatogr.,* 111, 413, 1975.

23. **Siezen, R. J. and Mague, T. H.,** Gas-liquid chromatography of the N-heptafluorobutyryl isobutyl esters of fifty biological interesting amino acids, *J. Chromatogr.,* 130, 151, 1977.

24. **Pearce, R. J.,** Amino acid analysis by gas-liquid chromatography of N-heptafluorobutyryl isobutyl esters. Complete resolution using a support-coated open-tubular capillary column, *J. Chromatogr.,* 136, 113, 1977.

25. **Felker, P.,** Gas-liquid chromatography of the heptafluorobutyryl O-isobutyl esters of amino acids, *J. Chromatogr.,* 153, 259, 1978.

26. **Desgres, J., Boisson, D., and Padieu, P.,** Gas-liquid chromatography of isobutyl ester, N(O)-heptafluorobutyrate derivatives of amino acids on a glass capillary column for quantitative separation in clinical biology, *J. Chromatogr.,* 162, 133, 1979.

27. **MacKenzie, S. L. and Tenaschuk, D.,** Quantitative formation of N(O,S)-heptafluorobutyryl isobutyl amino acids for gas chromatographic analysis. I. Esterification, *J. Chromatogr.,* 171, 195, 1979.

28. **Bengtsson, G. and Odham, G.,** A micromethod for the analysis of free amino acids by gas chromatography and its application to biological systems, *Anal. Biochem.,* 92, 426, 1979.

29. **Moodie, I. M.,** Gas-liquid chromatography of amino acids. The heptafluorobutyryl-isobutyl ester derivative of tryptophan, *J. Chromatogr.,* 208, 60, 1981.

30. **Chauhan, J., Darbre, A., and Cariyle, R. F.,** Determination of urinary amino acids by means of glass capillary gas-liquid chromatography with alkaline-flame ionisation detection and flame ionisation detection, *J. Chromatogr.,* 227, 305, 1982.

31. **Chauhan, J. and Darbre, A.,** Determination of amino acids by means of glass capillary gas-liquid chromatography with temperature-programmed electron-capture detection, *J. Chromatogr.,* 236, 151, 1982.

32. **Zanetta, J. P. and Vincendon, G.,** Gas-liquid chromatography of the N(O)-heptafluorobutyrates of the isoamyl esters of amino acids. I. Separation and quantitative determination of the constituent amino acids of proteins, *J. Chromatogr.,* 76, 91, 1973.

33. **Felker, P. and Bandurski, R. S.,** Quantitative gas-liquid chromatography and mass spectrometry of the N(O)-perfluorobutyryl-O-isoamyl derivatives of amino acids, *Anal. Chem.,* 67, 245, 1975.

34. **Frank, H., Nicholson, G. J., and Bayer, E.,** Enantiomer labelling, a method for the quantitative analysis of amino acids, *J. Chromatogr.,* 167, 187, 1978.

35. **Frank, H., Rettenmeier, A., Weicker, H., Nicholson, G. J., and Bayer, E.,** A new gas chromatographic method for determination of amino acids levels in human serum, *Clin. Chim. Acta,* 105, 201, 1980.

36. **Frank, H., Rettenmeier, A., Welcker, H., Nicholson, G. J., and Bayer, E.,** Determination of enantiomer-labelled amino acids in small volume of blood by gas chromatography, *Anal. Chem.,* 54, 715, 1982.

37. **Coulter, J. R. and Hann, C. S.,** A practical quantitative gas chromatographic analysis of amino acids using the n-propyl N-acetyl esters, *J. Chromatogr.,* 36, 42, 1968.

38. **Coulter, J. R. and Hann, C. S.,** Gas chromatography of amino acids, in *New Techniques in Amino Acid, Peptide, and Protein Analysis,* Niederwieser, A. and Pataki, G., Eds., Ann Arbor Science, Ann Arbor, Mich., 1971, 75.

39. **McGregor, R. F., Britin, G. M., and Sharon, M. S.,** Determination of urinary amino acids by gas chromatography, *Clin. Chim. Acta,* 48, 65, 1973.

40. **Adams, R. F.,** Determination of amino acid profiles in biological samples by gas chromatography, *J. Chromatogr.,* 95, 189, 1974.

41. **Adams, R. F., Vandemark, F. L., and Schmidt, G. J.,** Ultramicro GC determination of amino acids using glass open tubular columns and a nitrogen-selective detector, *J. Chromatogr., Sci.,* 15, 63, 1977.

42. **Hušek, P.,** Gas chromatography of cyclic amino acid derivatives. A useful alternative to esterification procedures, *J. Chromatogr.,* 234, 381, 1982.

43. **Makita, M., Yamamoto, S., Kono, M., Sakai, K., and Shiraishi, M.,** A simple and convenient method for gas chromatographic analysis of amino acids, *Chem. Ind. (London),* 355, 1975.

44. **Makita, M., Yamamoto, S., and Kono, M.,** Gas-liquid chromatographic analysis of protein amino acids as *N*-isobutyloxycarbonyl amino acid methyl esters, *J. Chromatogr.,* 120, 129, 1976.

45. **Makita, M., Yamamoto, S., and Kiyama, S.,** Improved gas-liquid chromatographic method for the determination of protein amino acids, *J. Chromatogr.,* 237, 279, 1982.

46. **Makita, M. and Yamamoto, S.,** Determination of amino acids in protein and peptide hydrolysates by gas-liquid chromatography, *Yakugaku Zasshi,* 96, 816, 1976.

47. **Makita, M. and Yamamoto, S.,** Determination of amino acids in serum by gas-liquid chromatography, *Yakugaku Zasshi,* 96, 777, 1976.

48. **Makita, M., Yamamoto, S., and Tsudaka, Y.,** Gas chromatographic determination of hydroxyproline in urine hydrolysates, *Clin. Chim. Acta,* 88, 305, 1978.

49. **Matsu-ura, S., Yamamoto, S., and Makita, M.,** Determination of γ-carboxyglutamic acid in human urine by gas-liquid chromatography, *Anal. Biochem.,* 114, 371, 1981.

50. **Yamamoto, S., Kiyama, S., Watanabe, Y., and Makita, M.,** Practical gas-liquid chromatographic method for the determination of amino acids in human serum, *J. Chromatogr.,* 233, 39, 1982.

51. **Fernlund, P., Stenflo, J., Roepstorff, P., and Thomsen, J.,** Vitamin K and the biosynthesis of prothrombin. V. γ-Carboxyglutamic acid, the vitamin K-dependent structures in prothrombin, *J. Biol. Chem.,* 250, 6125, 1975.

52. **Horning, E. C., VandenHeuvel, W. J. A., and Creech, B. G.,** Separation and determination of steroids by gas chromatography, *Methods Biochem. Anal.,* 11, 69, 1963.

53. **Makita, M., Yamamoto, S., Katoh, A., and Takashita, Y.,** Gas chromatography of some simple phenols as their *O*-isobutyloxycarbonyl derivatives, *J. Chromatogr.,* 147, 456, 1978.

54. **Makita, M., Yamamoto, S., Miyake, M., and Masamoto, K.,** Practical gas chromatographic method for the determination of urinary polyamines, *J. Chromatogr.,* 156, 340, 1978.

55. **Yamamoto, S., Kakuno, K., Okahara, S., Kataoka, H., and Makita, M.,** Gas chromatography of phenolic amines, 3-methoxycatecholamines, indoleamines and related amines as their *N,O*-ethyloxycarbonyl derivatives, *J. Chromatogr.,* 194, 399, 1980.

56. **Schlenk, H. and Gellerman, J. L.,** Esterification of fatty acids with diazomethane on a small scale, *Anal. Chem.,* 32, 1412, 1960.

57. **Walker, M. A., Roberts, D. R., and Dumbroff, E. B.,** Convenient apparatus for methylating small samples with diazomethane, *J. Chromatogr.,* 241, 390, 1982.

58. **Moodie, I. M.,** Gas-liquid chromatography of amino acids. A solution to the histidine problem, *J. Chromatogr.,* 99, 495, 1974.

59. **LeRoy, E. C.,** The technique and significance of hydroxyproline measurement in man, in *Advances in Clinical Chemistry,* Vol. 10, Bodansky, O. and Stewart, C. P., Eds., Academic Press, New York, 1967, 213.

60. **Fernlund, P.,** γ-Carboxyglutamic acid in human urine, *Clin. Chim. Acta,* 72, 147, 1976.

61. **Levy, R. J. and Lian, J. B.,** γ-Carboxyglutamate excretion and warfarin therapy, *Clin. Pharmacol. Ther.,* 25, 562, 1979.

62. **Makita, M., Yamamoto, S., Sakai, K., and Shiraishi, M.,** Gas-liquid chromatography of the *N*-isobutyloxycarbonyl methyl esters of non-protein amino acids, *J. Chromatogr.,* 124, 92, 1976.

Chapter 7

CAPILLARY GLC OF AMINO ACIDS*

Robert W. Zumwalt, Jean Desgres, Kenneth C. Kuo, James E. Pautz, and Charles W. Gehrke

TABLE OF CONTENTS

*   Reproduced with permission from the *Journal of the Association of Official Analytical Chemists.*

I. INTRODUCTION

Separation and measurement of amino acids as their volatile derivatives has mainly been performed with conventional packed gas chromatography (GC) columns with two major exceptions. The first is the use of wall-coated open tubular (WCOT) (capillary) columns for separation of amino acid diastereomers or enantiomers, and the second is the use of capillary columns interfaced with mass spectrometry (MS) for structural elucidation and identification of amino acids in very complex matrices (e.g., physiological fluids) which contain numerous nonprotein amino acids as well as the protein amino acids.

The large numbers of theoretical plates which can be obtained by use of long, narrow-bore capillary columns was demonstrated by Golay[1] in 1958. Although the number of theoretical plates per unit length may be similar for packed and capillary columns, the low resistance to carrier gas flow in capillary columns permits use of much longer columns, and thus the total number of plates provided by capillary columns is much higher.

However, capillary gas-liquid chromatography (GLC) columns constructed of stainless steel or glass did not achieve the widespread, routine use which would seem to be warranted by their superior efficiency as compared to packed columns. Difficulties in coating capillary columns with a thin, uniform layer of stationary phase, the brittleness of glass capillaries, and the somewhat more complicated technical arrangements (inlet systems and makeup gas) hampered the routine use of capillary GLC for amino acid analysis. Lipsky[2] has recently provided an excellent review of the history of the development of capillary columns.

As MacKenzie has so clearly pointed out in Chapter 4, Volume I, the evolution of amino acid analysis constantly demands improvements in resolution, sensitivity, precision, and accuracy, along with reduced capital and operating costs. Although the well-established and automated ion exchange amino acid analyzers meet many demands, the methodology suffers from some limitations. MacKenzie notes that expensive, dedicated equipment which is not quickly adaptable to other analyses is required, and that ion exchange amino acid analysis is somewhat inflexible in responding to specific analytical problems. Although new resins are shortening the times required for protein hydrolysate and physiological sample analyses, the times required for analyses generally remain substantial. For example, our long-term experience is that the Beckman® 121 M amino acid analyzer with Benson® BX8 resin (8 ± 1 μm diameter) requires 130 min for a hydrolysate analysis and 240 min for a physiological sample. Further, the successful operation of an ion-exchange instrument requires the dedication of a trained and skilled technician. Perhaps the most serious limitation to the ion-exchange technique, as MacKenzie has also noted, is the lack of an interface to a mass spectrometer so that compound identification can be based on a more absolute criterion than retention time alone. With reference to all of the above, GLC provides the answers to these problems, and also has the method attributes of selectivity, sensitivity, precision, accuracy, economics, and versatility.

In 1966, Gil-Av et al.[3] reported the first separation of amino acid enantiomers by GLC with an optically active stationary phase. Using *N*-trifluoroacetyl (TFA)-L-isoleucine lauryl ester as the stationary phase for a 100-m glass capillary column, he achieved the separation of a number of amino acid enantiomers. In Chapter 1 of Volume II, Gil-Av provides a thorough discussion of the development of chiral diamide stationary phases for the resolution of amino acids.

In 1977, Frank et al[4] synthesized a novel-type chiral stationary phase which possessed greatly enhanced thermal stability. The stationary phase, Chirasil-Val®, is a polysiloxane functionalized with carboxyl alkyl groups to which valine-*t*-butylamide residues are coupled through an amide linkage. Synthesis of this phase has led to Bayer's development of the separation of amino acid enantiomers using chiral polysiloxanes and the novel technique of quantitative amino acid analysis by "enantiomeric labeling." As Bayer, Gil-Av, and others have pointed out in Volumes II and III, Chirasil-Val® or similar diamide phases can resolve enantiomers of numerous classes including amines, vicinal amino alcohols, α-hydroxycarboxylic acids, glycols, sulfoxides, alcohols, and carbohydrates. The separation and analysis of diastereomers represents another important contribution which capillary GC offers, and König[5] provides an excellent discussion of the area in this volume.

The value of capillary GC for the amino acid analysis of extremely complex samples is clearly illustrated by Desgres and Padieu in Chapter 5, Volume I in their research in the area of clinical biology. As they point out, there are about 50 diseases known to be due to anomalies of amino acid metabolism, with most resulting in mental retardation. Although partial or total prevention of the effects can be accomplished by dietary means, it is essential to establish a diagnosis as early as possible by means of analysis of the free amino acids in physiological fluids. They state that the analytical technique for such assays must not only be specific, sensitive, and precise, but also sufficiently rapid to permit daily execution of several assays of the same case, and that GC and GC/MS are perfectly matched to these criteria. In their chapter, Desgres and Padieu provide an excellent description of the value and use of capillary GC for detecting and investigating metabolic diseases, and conclude that " . . . GC has proven to be well adapted to clinical chemistry due to the low cost of equipment and sound methodology. Capillary GLC is easily dedicated not only to routine work but also to crucial determinations in the field of inborn metabolic disorders where rapid, precise, and repetitive analyses will greatly help the clinician to prevent dramatic and definitive injury among newborn babies."

The use of capillary GC to separate a large number of compounds in biological samples for pattern recognition has been described by Jellum et al.[6] in which over 100 peaks are used to classify cells as to their diseased state.

In 1985, Schneider et al.[7] used capillary GC to compare profiles of amino acids in urine of healthy individuals with amino acid profiles in hemofiltrates from uremic patients. They showed that hemofiltration removes some amino acids to a much greater extent from uremic patients than the kidney does from healthy persons. These researchers reported that hemofiltration performed three times a week removes 10 times the amount of methionine and 40 times the amount of leucine as compared to their excretion in urine by a healthy individual over the same period. Using the N-TFA n-butyl ester and N-HFB (heptafluorobutyryl) n-butyl ester derivatives, they reported that GC methods were very useful for detecting previously unknown compounds and for obtaining an overview of metabolic states. Since amino acids are involved in many metabolic diseases, the method may be used to attain a better insight into these diseases and to aid diagnosis. Their finding that large amounts of proline, lysine, leucine, and methionine are lost in the course of hemofiltration was not previously known.

Although the use of capillary GC has expanded greatly to encompass a wide range of analytical tasks involving many types of organic molecular classes, Gil-Av's statement[3] in 1966 that " . . . the achievement of resolution of optical isomers is one of the most striking demonstrations of the efficiency of gas liquid partition chromatography" is as appropriate now as it was 20 years ago.

Two recent developments have very significantly allowed major advancements to be made in using capillary GC, the result being much more widespread use of capillary GC in investigations on a broad range of chemical and biological problems. The two technological developments were the introduction of fused silica capillary columns by Dandeneau and Zerenner[8] in 1979, and the development by 1981 of immobilized stationary phases for capillary GC columns by a number of investigators.[2] Fused silica columns are produced from amorphous silica which is much lower in metallic impurities than naturally occurring quartz; the columns possess very low surface activity, and are inherently straight.

The immobilized liquid phases cannot be solvent-extracted from the column, but the stationary phase still shows all the characteristics of a liquid. This can be achieved by techniques such as the synthesis of the organic layer on the solid surface or by using the regular coating techniques to deposit a liquid stationary layer which is then immobilized by formation of covalent bonds between the polymer molecules (cross-linking) as well as between the molecules and the column wall itself (bonding). The covalent bonds are induced by the use of peroxides dissolved in the stationary phase. Grob et al.[9,10] and others[2] have published comprehensive papers on stationary phase immobilization. As fused silica columns with immobilized stationary phases of varying polarities are offered by numerous vendors of chromatographic equipment and supplies, they have become widely used for numerous analytical tasks.

Labadarios et al.[11] have observed in a critical review that despite the added time required for derivatization, GC analysis of amino acids offers advantages in capital and operating costs, shorter elution time, and improved precision and accuracy as compared to amino acid analyzers.

The objective of our study was to investigate the applicability of commercially available fused silica capillary columns for their effectiveness in separation of amino acids and quantitation on comparison with the classical ion exchange method.

We selected the *N*-TFA *n*-butyl and the *N*-HFB isobutyl ester derivatives for this study because of the extensive research and application of these derivatives during the past 20 years. The amino acid content of hydrolysates of five diverse materials was measured: ribonuclease, β-lactoglobulin, lysozyme, soybean meal, and a commercial poultry feed. Single 6 *N* HCl hydrolysates of each material were prepared to minimize sample preparation differences, and three independent analyses of each hydrolysate were made by each of three techniques; the *N*-TFA *n*-butyl and *N*-HFB isobutyl ester methods using capillary GC, and by ion exchange chromatography (IEC) with a Beckman® 121 M amino acid analyzer.

Our results clearly demonstrate that capillary GC analysis of amino acids using fused silica bonded phase columns provides data with good precision and in generally excellent agreement with ion exchange analyses. With simply the purchase of a commercially available column and the appropriate reagents, any laboratory with a capillary GC capability can perform quantitative measurements of amino acids.

II. EXPERIMENTAL

A. Apparatus

All analyses were performed with a Hewlett-Packard® 5880 Gas Chromatograph with a Level Four Data System. The GC was equipped with a flame ionization detector (FID) and an on-column and split injection systems (HP-1932OH). Helium, hydrogen, and air were "zero grade" (Linde Specialty Gases, a subsidiary of Union Carbide, Somerset, N.J.). An Oxisorb® (Supelco, Inc., Bellefonte, Pa.) oxygen scrubber was placed in line with the helium carrier gas. Durabond® fused silica capillary columns were obtained from J&W Scientific, Rancho Cordova, Calif., the DB-1 (methyl silicone) column was 20 m × 0.25 mm I.D., and the DB-5 (5% phenyl methyl silicone) column was 30 m × 0.32 mm I.D., both had 0.25 μm film thickness.

The aqueous amino acid solutions were evaporated in a Speed Vac® concentrator (Savant Instruments, Inc. Hicksville, N.Y.) at reduced pressure or under a nitrogen stream at 55°C. The sample tubes (13 × 100 mm or 16 × 75 mm) were glass with PTFE-lined screw caps (Corning® Glassware). For esterification and acylation, the samples were heated in a TEMP-BLOK® module heater (Lab-Line Instruments, Inc., Melrose Park, Ill.). Injections were made using a 701SN, 10-μℓ Hamilton® (Hamilton, Reno, Nev.) on-column syringe with a 12.5 cm × 0.24 mm I.D., 32 ga. fused silica needle.

B. Reagents

Trifluoroacetic anhydride (TFAA) and heptafluorobutyric anhydride (HFBA), were purchased from Regis Chemical Company, Morton Grove, Ill. TFAA and HFBA were also obtained from Pierce Chemicals (Rockford, Ill.) and Aldrich Chemicals (Milwaukee, Wisc.). The isobutanol · 3 *N* HCl and the *n*-butanol · 3 *N* HCl were prepared by adding 6.75 mℓ of distilled acetyl chloride (Aldrich Chemicals Gold label Reagent) to 25 mℓ of isobutanol or *n*-butanol at 0°C. All reagents were stored at 4°C until used. Nanograde dichloromethane was obtained from Mallinckrodt Chemical Co., Paris, Ky. Ethyl acetate was purchased from Burdick & Jackson High Purity Solvents, Muskegon, Mich. Molecular sieve pellets were added to ethyl acetate and dichloromethane and allowed to stand overnight to absorb water. GLC analysis of a performance blank indicated no background peaks.

C. *N*-TFA *n*-Butyl Ester Derivatives

A stock solution of amino acids (0.10 mg of each amino acid per milliliter in 0.1 *N* HCl) was prepared and 200 μℓ were pipetted into a sample tube. To this solution, 200 μℓ of a norleucine internal standard (0.10 mg/mℓ in 0.1 *N* HCl) were added. The sample was then evaporated to dryness under a gentle stream of nitrogen at 55°C. The *n*-butanol · 3 *N* HCl was allowed to warm to ambient temperature, and 600 μℓ were added to the sample. The sample tube was capped and mixed by sonication for about 10 sec. The sample tube was then placed in a 100 ± 1°C heating block containing silicone oil to the level of liquid inside the tube. After heating for 30 min, the sample tube was removed from the heat block and allowed to cool to room temperature. Dry nitrogen gas was used to evaporate the *n*-butanol · 3 *N* HCl and the sample was twice azeotropically dried with 0.5 mℓ of dichloromethane with N$_2$ sweep at room temperature. A 1:2 mixture of TFAA/CH$_2$Cl$_2$ was prepared and 1.00 mℓ was added to the dried sample and heated as before at 150°C for 5 min. When cool, approximately 0.5 to 0.75 μℓ was injected on-column.

D. *N(O)*-HFB Isobutyl Esters

The standard amino acid solution mixture previously described was derivatized by placing 200 μℓ of the standard solution and 200 μℓ of norleucine solution in a sample tube and evaporating to dryness under a nitrogen stream at 55°C. To the dried sample, 200 μℓ of isobutanol · 3 *N* HCl were added. Then the sample tube was capped and heated for 45 min in a 110 ± 1°C heating block. After cooling, isobutanol · 3 *N* HCl was evaporated to dryness under a nitrogen stream at 40°C. Then 80 μℓ of ethyl acetate and 20 μℓ of HFBA were added and the mixture was heated for 20 min in the heating block at 110°C. After cooling, the solution was ready for GC analysis; if needed, the sample can be diluted with ethyl acetate.

The relative weight responses of each amino acid relative to the internal standard norleucine was determined as both the *N*-TFA *n*-butyl and *N*-HFB isobutyl derivatives using a standard reference solution of amino acids. These data are given in Table 1 and were used to compute the concentration of each amino acid in the sample hydrolysates.

E. Hydrolysate Preparation

A single hydrolysate of each of five proteinaceous materials (ribonuclease, β-lactoglobulin, lysozyme, soybean meal, poultry feed) was prepared. A quantity of each sample corresponding to 20 mg of protein was weighed into a Pyrex® glass hydrolysis tube. The hydrolysis tubes (205 × 20 mm) had two openings, one (female-threaded) at the top and a side arm 45 mm from the top of the tube. The tube could be sealed both above and below the side arm with a male-threaded plunger. By backing the plunger part way out of the tube, the seal below the side arm could be broken, allowing removal of air via the side arm, while the top of the tube remained sealed.

After transfer of the samples into the tubes, 10.0 mℓ of 6 *N* HCl were added and the samples mixed by Vortex® mixer. The samples were then frozen in a dry ice/acetone bath, and the sidearms attached to a vacuum pump for removal of air. After reducing the pressure inside the tube to 200 μm Hg, the side arms were sealed with the plunger, removed from the dry ice/acetone bath, and allowed to warm to room temperature for degassing of the 6 *N* HCl. This freezing-air evacuation-thawing procedure was performed three times on each sample to minimize the presence of air during hydrolysis. The samples were then placed in a 145°C heating block for 4 hr. After hydrolysis, the samples were transferred to 50 mℓ volumetric flasks and brought to volume with nanopure water. Aliquots were taken for analysis, and the remaining hydrolysates were transferred to glass bottles with Teflon®-lined screw caps and stored at -20°C.

The preparation of hydrolysates is a critical step in determining the amino acid composition of proteinaceous materials. Our recent study on sample preparation and hydrolysis[12] demonstrated that inaccuracies and lack of precision which arise are mainly due to sample handling and the hydrolysis technique rather than the method of analysis. Ion exchange and GC both provide excellent precision and accuracy.

III. COMMENTS ON THE METHOD

A. Instrumental

1.   Gases of the highest purity must be used at all times and a trap consisting of activated carbon, indicating calcium sulfate, and the molecular sieve must be placed in the line of each of the air, helium, and hydrogen tanks. Gas regulators with rubber diaphragms should be avoided since these diaphragms can cause contamination of the gas flow system causing baseline noise as well as deterioration of the column. An oxygen scrubber should also be placed in line with the carrier gas and changed at least every other tank. The tank should be changed before the pressure drops below 500 psi. As the gas is used, the lighter molecules escape out the regulator by effusion and thus only higher molecular weight impurities concentrate in the remaining gas in the tank. Once inside the GC, impurities are very difficult and time-consuming to remove.

2.   Capillary columns are very durable and flexible, but care must always be used during handling. The cutting of capillary columns is best performed with a carbide-tipped scoring pencil. Lightly abrading the polyimide coating but not breaking the fused silica and bending the column with the scored portion down will effect a clean square break. The column should be held with the newly broken ends pointing down so that the particles from the break do not fall into the column.

3.   Graphite ferrules are used to seal the column to the injector and detector inlets. It is best to first run the column through the ferrule then break off a few centimeters to ensure no graphite is within the column. Keeping these ends clean saves time in conditioning the column.

4.   On-column injectors such as Hewlett-Packard's use a duck-bill valve to isolate the carrier gas from the atmosphere. A specified minimum head pressure of at least 30 psi must be used to properly seal this valve. Variations in retention times and gas escaping from the top of the injector are symptomatic of incomplete sealing of the duck-bill valve.

5.   Alignment problems with the syringe and column can be avoided by placing a small gauge cleaning wire down the injector, and sliding the column up onto the wire. Insert the column into the injector until it stops, then begin threading the backup nut into the injector but do not tighten it. Withdraw the column downward about 0.5 mm and tighten the backup nut, then withdraw the wire from the injector.

6.   The detector end of the column should be slowly pushed into the jet tip. When it stops, start threading the backup nut but do not tighten it. Withdraw the column about 1 mm and tighten the backup nut gently about $^1/_4$ turn. Over tightening either end of the column will crush the capillary. Leaks can be checked with a liquid leak detector.

B. Conditioning and Storage of Columns

Bonded-phase columns are generally preconditioned by the manufacturer and for most applications do not require extensive further conditioning.

1. After installation of the column in the instrument, purge the air from the system before heating the column. Even trace amounts of air can seriously damage the column at elevated temperatures.
2. Note the maximum temperature limit of the column, as in no case should the column be exposed to temperatures above its maximum temperature limit.
3. The column should only be programmed to, or slightly above, your highest operating temperature and only for a short period of time. Hold this temperature just long enough to achieve a stable (or perhaps slightly falling) baseline. No further conditioning is required for most applications.
4. The more polar the column, and the longer the storage period, the more important it is to seal the column ends during storage. For long-term storage, flame sealing is advisable. For short-term storage or storage of less polar column, forcing a septum on each end of the column is generally satifactory. Be sure to remove a few cm of column, after installing the ferrule, prior to re-installation.

C. Methodology

1. Amino acid solutions are kept refrigerated or frozen in $0.1\ N$ HCl purged with N_2.
2. Esterification alcohols should be kept as anhydrous as possible. This is difficult since HCl causes dehydration of the alcohol to the corresponding alkenes and water. Another possible reaction is the nucleophillic attack of chloride ion on the protonated alcohol forming chloroalkanes and water. To minimize these side reactions, it is best to keep the esterifying alcohols refrigerated. Always allow them to reach room temperature before use.
3. Careful removal of water formed in the esterification step and excess alcohol is essential for reproducible and accurate analyses. Adding two 0.5-mℓ aliquots of dry methylene chloride and drying with N_2 sweep is sufficient for this purpose.
4. TFAA and HFBA react violently with water. They are corrosive to skin and mucous membranes. Always use them in a well ventilated area and avoid contact with the skin.
5. Derivatized samples should be chromatographed as soon as possible. If samples must be stored, they should be kept under refrigeration.

1. Example Calculation

Analysis of an equal weight amino acid standard reference solution is used to calculate the relative weight response (RWR) of each amino acid to the internal standard (IS) norleucine.

$$RWR_{AA/IS} = \frac{\text{Peak Area of Amino Acid/Unit Weight}}{\text{Peak Area of Internal Standard/Unit Weight}}$$

For example, the RWR for alanine as the *N*-TFA *n*-butyl ester is seen to be 0.965 in Table 1.

The amount of each amino acid in the sample, expressed as w/w%, is calculated as follows:

$$Q_{AA} = \frac{1}{RWR_{AA/IS}\ \text{Std.}} \times \frac{\text{Area}_{AA}}{\text{Area}_{IS}\ \text{Sample}} \times Q_{IS}/\text{Sample Weight} \times 100$$

Table 1

**RELATIVE WEIGHT RESPONSES OF AMINO
ACID *N*-TFA *n*-BUTYL AND *N*-HFB ISOBUTYL
ESTERS[a]**

	N-TFA *n*-Butyl			*N*-HFB Isobutyl		
	x^b	S.D.	%RSD	x^b	S.D.	%RSD
Ala	0.965	0.012	1.2	1.015	0.001	0.1
Gly	0.906	0.009	1.0	1.042	0.004	0.4
Thr	0.867	0.007	0.8	0.978	0.013	1.3
Ser	0.841	0.012	1.4	0.961	0.015	1.6
Val	0.997	0.015	1.5	0.981	0.030	3.1
Leu	1.001	0.006	0.6	0.992	0.007	0.7
Ile	0.947	0.057	6.0	0.924	0.060	6.5
Pro	1.012	0.008	0.8	1.020	0.008	0.8
Met	0.730	0.055	7.5	0.703	0.049	7.0
Asp	0.923	0.024	2.6	1.043	0.016	1.5
Phe	1.193	0.023	1.9	1.061	0.008	0.8
His	0.288	0.028	9.7	0.328	0.050	15.1
Tyr	0.963	0.010	1.0	1.026	0.020	2.0
Glu	1.047	0.007	0.7	1.075	0.017	1.6
Lys	0.842	0.018	2.1	1.015	0.026	2.6
Arg	0.655	0.066	10.1	0.794	0.063	8.0
Cys	0.508	0.024	4.7	0.747	0.024	3.3

[a]   Relative to norleucine as the internal standard.
[b]   Mean of three independent analyses of a single amino acid standard
solution.

Q_{AA} is the w/w% of amino acid in the sample; Q_{IS} is the amount of internal standard added
to the sample

To illustrate, a 20.0-mg sample is hydrolyzed and brought to a volume of 50.0 mℓ. Then,
1.00 mℓ of the diluted hydrolysate is taken (0.400 mg), 0.020 mg of internal standard is
added, and the sample is taken through the derivatization procedure and analyzed. Peak
areas found: Ala, 31308, IS, 21656.

$$Q_{AA} = \frac{1}{0.965} \times \frac{31308}{21656} \times 0.020 \text{ mgIS}/0.400 \text{ mg Sample} \times 100$$

$$= 1.036 \times 1.446 \times 0.020 \text{ mgIS}/0.400 \text{ mg Sample} \times 100$$

$$= 0.075 \times 100$$

$$Q_{ALA} = 7.50\%$$

Thus this sample was found to contain 7.50% alanine.

Many gas chromatographs are equipped with computer devices which will perform these
calculations and directly print the results. For example the HP-5880 allows the choice of
four computation methods; area%, normalized area%, and external and internal standard
methods.

IV. RESULTS AND DISCUSSION

Fused silica capillary GC analyses of amino acids were performed using both the *N(O)*-
TFA *n*-butyl derivatives and the *N(O)*-HFB isobutyl derivatives. IEC analyses were per-

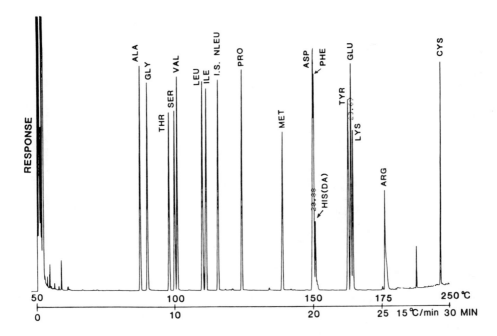

FIGURE 1. Fused silica capillary GC analysis of amino acids *N(O)*-TFA *n*-butyl esters. Standard solution, final sample conc.: 400 μg/1.0 mℓ; injection: on-column, ~15 ng each amino acid (0.75 μℓ); column: 30 m × 0.32 mm I.D., 0.25 μℓ film thickness; initial temp.: 50°C, 5°C, 5°/min to 175°, DB-5 (phenylmethyl silicone) 15°/min to 250°C.

formed with a Beckman® 121 M Amino Acid Analyzer. All analyses were performed in triplicate on the same five hydrolysates.

A. *N(O)*-TFA *n*-Butyl Esters

Figure 1 shows the separation of the TFA *n*-butyl derivatives on a 30 m, 0.32 mm I.D., DB-5 (10% phenyl, methyl silicone) column, 0.25 μm film thicknesses, with aspartic acid, phenylalanine, and histidine being the amino acids most difficult to separate. Use of 40- and 60-m DB-5 columns with 1.0-μm films resulted in nonelution of the cystine derivative. Investigation of a less polar column (DB-1, methyl silicone), and two columns more polar than DB-5 (DB-17, 50% phenyl, methyl silicone, and RLS-400, polytrifluoropropyl siloxane) resulted in either nonseparation of all the TFA *n*-butyl derivatives, nonelution of cystine, or poor peak shape for some amino acids. Use of DB-5 column with a smaller I.D., 0.25 mm, gave improved resolution of aspartic acid and phenylalanine, and also of tyrosine-glutamic acid-lysine as seen in Figure 2. However, we found on-column injection into 0.25-mm columns to be tedious and time-consuming, and not suited for routine analysis. Investigations to obtain the complete separation and elution of the *N*-TFA *n*-butyl derivatives with fused silica columns are continuing. However, the DB-5 column is effective for analyzing the protein amino acids other than Asp, Phe, and His as seen in Figures 2 and 3 and Tables 2 to 4. The analysis of the lysozyme hydrolysate is shown Figure 3, and Figure 4 shows the chromatogram obtained from the ribonuclease hydrolysate. The two nonpure protein hydrolysates (soybean meal and poultry feed) did not require ion exchange cleanup or any sample purification procedure prior to derivatization and chromatography. However, it is true that with more complex matrices as blood serum an ion exchange cleanup would be necessary as described by Gehrke, Bayer, Desgres, and others in these volumes to obtain reliable data. The *N*-TFA *n*-butyl esters were analyzed using the on-column injection technique, as we have found analysis of arginine and cystine more reproducible with this type of injection.

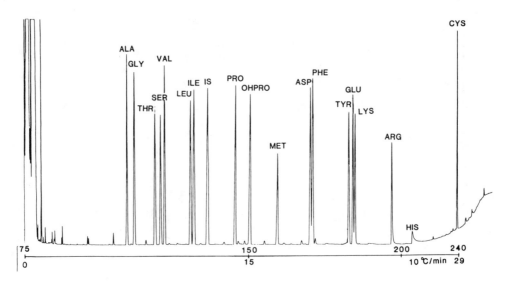

FIGURE 2. Bonded phase fused silica capillary GLC of amino acid *N*-TFA *n*-butyl esters. Standard solution, final sample conc.: 400 μg/1.0 mℓ; injection: on-column, ~15 ng each amino acid (0.75 μℓ); column: 0.25 mm I.D., 0.25 μm film thickness; initial temp.: 75°C, 5°/min to 200°, DB-5(phenylmethyl silicone) 20 m 10°/min to 240°C.

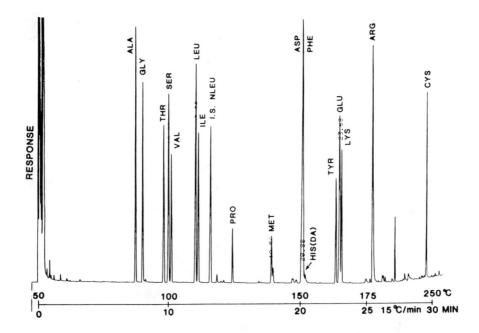

FIGURE 3. Fused silica capillary GC analysis of amino acids *N(O)*-TFA *n*-butyl esters. Lysozyme hydrolysate, final sample conc.: 400 μg/1.0 mℓ; injection: on-column, ~15 ng each amino acid (0.75 μℓ); column: 30 m × 0.32 mm I.D., 0.25 μm film thickness; initial temp.: 50°C, 5°/min to 175°, DB-5 (phenylmethyl silicone) 15°/min to 250°C.

B. *N(O)*-HFB Isobutyl Esters

Analysis of a standard solution of amino acids as the HFB isobutyl derivatives is shown in Figure 5. A 20-m DB-1 (methyl silicone) column separates the protein amino acids, and Figure 6 shows the separation of amino acids in a "collagen standard" which is a commercially obtained amino acid standard solution for analysis of collagen samples. This

Table 2
COMPARISON OF FUSED SILICA CAPILLARY GC AND IEC ANALYSES OF PROTEIN HYDROLYSATES[a,b]

	Ribonuclease					Lysozyme				
	w/w%			Rel% diff.[c]		w/w%			Rel% diff.	
	N-TFA n-Butyl	N-HFB iso-Butyl	IEC	TFA	HFB	N-TFA n-Butyl	N-HFB iso-Butyl	IEC	TFA	HFB
Ala	6.98	6.96	6.84	2.05	1.75	7.08	7.24	6.93	2.16	4.47
Gly	1.56	1.56	1.59	1.89	1.89	5.99	6.13	5.87	2.04	4.42
Thr	7.47	7.24	7.28	2.61	0.55	5.33	5.42	5.29	0.76	2.46
Ser	9.54	9.43	9.11	4.72	3.51	6.57	6.82	6.59	0.30	3.49
Val	6.29	6.13	5.93	5.56	3.37	3.69	3.65	3.47	6.34	5.19
Leu	1.80	1.75	1.75	2.86	0.00	6.67	6.74	6.64	0.45	1.51
Ile	1.55	1.43	1.61	3.73	11.18	4.52	4.21	4.08	10.78	3.19
Pro	3.07	3.18	3.06	0.33	3.92	1.50	1.54	1.46	2.74	5.48
Met	3.17	2.98	3.54	10.45	15.82	1.83	1.68	1.97	7.11	14.72
Asp	NR[d]	11.31	12.74	NR	11.22	NR	16.27	17.86	NR	8.90
Phe	NR	3.10	3.42	NR	9.36	NR	3.05	3.30	NR	7.58
His	NR	4.88	3.75	NR	30.13	NR	1.13	1.08	NR	4.63
Tyr	6.04	5.99	6.16	1.95	2.76	3.49	3.53	3.69	5.42	4.34
Glu	11.44	11.04	11.53	0.78	4.29	4.71	4.79	4.84	2.69	1.03
Lys	9.03	9.81	9.63	6.23	1.87	5.22	5.27	5.42	3.69	2.77
Arg	4.76	4.52	4.48	6.25	0.89	10.88	11.44	12.21	10.89	6.31
Cys	4.53	4.95	4.84	6.40	2.27	5.26	6.30	5.26	0.00	19.77

[a] Each value represents the mean of three analyses of a single hydrolysate.
[b] Samples hydrolyzed with 6 N HCl, 145°-4 hr, under vacuum.
[c] Relative % differences are the GC values/IEC values × 100.
[d] Asp, Phe, and His peaks not resolved (NR).

solution contains elevated amounts of proline and hydroxyproline, plus D,L- and *allo*-hydroxylysine which are separated and designated HYLYS on the chromatogram. In the HFB isobutyl ester procedure, the acylating solution was removed by evaporation and the derivatives dissolved in the ethyl acetate prior to injection which produces both diacyl and nonoacyl histidine derivatives. Desgres and Padieu describe in Chapter 5, Volume I, the analysis of histidine after removal of the acylating solution and reaction of the HFB isobutyl histidine derivative with diethoxyformic anhydride to attach the ethoxyformic group to a ring nitrogen of histidine. The analysis of ribonuclease is presented in Figure 7.

C. Analytical Results

The data from the three analytical methods are shown in Tables 2 to 4, and in summary there is generally excellent agreement among the three methods for all five hydrolysates. The data show that none of the methods gave consistently high or low results in comparison with the other two methods. As a measure of the differences between the GC and IEC values, relative percent differences were calculated for each amino acids.

The relative percent differences between the GLC and ion exchange revealed no bias with regard to the method of analysis. For individual amino acids, GLC gave values greater than ion exchange in 75 instances, while the reverse was true in 71 instances. For ribonuclease, the differences for the TFA *n*-butyl method as compared to IEC ranged from 0.33% for proline to 10.45% for methionine with a mean relative percent difference of 14 amino acids of 3.99%. The HFB isobutyl method gave very similar results with a mean relative percent difference of 3.86% for the same 14 amino acids. Agreement among the two sets of GC

Table 3

**COMPARISON OF FUSED SILICA CAPILLARY GC AND IEC
ANALYSES OF PROTEIN HYDROLYSATES[a,b]**

| | β-Lactoglobulin | | | | | Soybean meal | | | | |
| | w/w% | | | Rel% diff.[c] | | w/w% | | | Rel% diff. | |
	N-TFA n-Butyl	N-HFB iso-Butyl	IEC	TFA	HFB	N-TFA n-Butyl	N-HFB iso-Butyl	IEC	TFA	HFB
Ala	5.57	5.60	5.50	1.27	1.82	1.93	2.11	2.03	4.93	3.94
Gly	0.97	1.04	1.04	6.73	0.00	1.95	2.08	2.00	2.50	4.00
Thr	4.06	4.02	4.05	0.25	0.74	1.82	1.92	1.86	2.15	3.23
Ser	3.12	3.19	3.06	1.96	4.25	2.52	2.59	2.52	0.00	2.78
Val	4.47	4.54	4.32	3.47	5.09	1.99	2.09	1.98	0.51	5.56
Leu	12.46	12.69	12.36	0.81	2.67	3.47	3.64	3.56	2.53	2.25
Ile	4.45	4.75	4.56	2.41	5.26	1.85	2.03	2.02	3.47	0.50
Pro	4.12	4.13	4.06	1.48	1.72	2.29	2.44	2.36	2.97	3.39
Met	2.41	2.33	2.57	6.23	9.34	0.70	0.55	0.64	9.38	14.06
Asp	NR[d]	8.47	9.43	NR	10.18	NR	5.02	5.39	NR	6.86
Phe	NR	2.74	3.00	NR	8.67	NR	2.36	2.37	NR	0.42
His	NR	1.85	1.26	NR	46.82	NR	1.63	1.25	NR	30.40
Tyr	3.03	2.83	3.19	5.02	11.29	1.59	1.69	1.74	8.62	2.92
Glu	16.73	16.04	16.46	1.64	2.55	8.45	8.74	8.70	2.87	0.46
Lys	9.62	9.59	9.51	1.16	0.84	2.62	3.05	2.93	10.58	4.10
Arg	2.07	2.03	2.03	1.97	0.00	3.26	2.81	3.42	4.55	17.83
Cys	2.24	2.23	2.21	1.36	0.90	0.47	0.52	0.56	16.07	7.14

[a]   Each value represents the mean of three analyses of a single hydrolysate.
[b]   Samples hydrolyzed with 6 N-HCl, 145°-4 hr, under vacuum.
[c]   Relative % differences are the GC values IEC values × 100.
[d]   Asp, Phe, and His peaks not resolved (NR).

and the IEC data was found for the lysozyme hydrolysate as seen in Table 2. Again the mean relative percent differences were small; 3.95% for the TFA *n*-butyl methods and 4.24% for the HFB isobutyl method. The HFB isobutyl value for cystine was considered and outlier was deleted from the mean value. Analyses of the β-lactoglobulin hydrolysate were likewise in very close agreement with mean relative percent differences of 2.55 and 3.32 % for the TFA *n*-butyl and HFB isobutyl methods as compared to the IEC data. The soybean meal and poultry feed hydrolysates contained some particulate residue after 6 *N* HCl hydrolysis. This material was removed by centrifugation of aliquots of the hydrolysates prior to evaporation and derivatization. No other sample cleanup was employed. Data from analyses of the soybean hydrolysate are shown in Table 3, with again very good agreement among the values obtained. The mean relative percent differences were 4.24% for the TFA *n*-butyl method and 4.17% for the HFB isobutyl method as compared to the IEC values. The poultry feed analyses are presented in Table 4, with mean relative percent differences of 4.24 and 4.17 for the TFA *n*-butyl and HFB isobutyl methods, respectively.

To assess which individual amino acids throughout the sample set might be most susceptible to GC-IEC analytical differences, the mean relative percent difference were calculated for individual amino acids (Table 5). Excellent agreement was found throughout the sample set for Ala, Gly, Thr, Ser, Val, Leu, Ile, Pro, Tyr, Glu, Lys, and Arg, as the TFA *n*-butyl esters. Phe, Asp, and His were unresolved. Methionine gave a mean relative percent difference of 10.63%, with the TFA *n*-butyl values being higher than the IEC values in 2 cases (soybean meal and poultry feed) and lower in the other 3 cases. For cystine, the TFA *n*-butyl and IEC values were in agreement for ribonuclease, lysozyme, and β-lactoglobulin,

Table 4
COMPARISON OF FUSED SILICA
CAPILLARY GC AND IEC ANALYSES OF
PROTEIN HYDROLYSATES[a,b]

	Poultry feed				
	w/w%			Rel% diff.[c]	
	N-TFA *n*-Butyl	*N*-HFB *iso*-Butyl	IEC	TFA	HFB
Ala	1.01	0.99	0.98	3.06	1.02
Gly	0.82	0.83	0.83	1.20	0.00
Thr	0.77	0.74	0.75	2.67	1.33
Ser	1.10	1.04	1.05	4.76	0.95
Val	0.79	0.76	0.72	9.72	5.56
Leu	1.66	1.62	1.62	2.47	0.00
Ile	0.68	0.68	0.72	5.56	5.56
Pro	1.18	1.20	1.16	1.72	3.45
Met	0.36	0.26	0.30	20.00	13.33
Asp	NR[d]	1.80	2.20	NR	10.89
Phe	NR	0.90	0.94	NR	4.26
His	NR	0.78	0.52	NR	50.00
Tyr	0.66	0.63	0.70	5.71	10.00
Glu	3.60	3.43	3.60	0.00	4.72
Lys	1.05	1.11	1.14	7.89	2.63
Arg	1.29	1.02	1.29	0.00	20.93
Cys	0.17	0.21	0.23	26.09	8.70

[a] Each value represents the mean of three analyses of a single hydrolysate.
[b] Samples hydrolyzed with 6 *N* HCl, 145°-4 hr, under vacuum.
[c] Relative % differences are the GC values/IEC values × 100.
[d] Asp, Phe, and His peaks not resolved (NR).

but varied by 16 and 26%, respectively for soybean meal and poultry feed, which contain only a fraction of the amount of cystine that the other three samples contain (0.2 to 0.5% as compared to 2 to 5%).

The HFB isobutyl values were generally in excellent agreement with the IEC values; Ala, Gly, Thr, Ser, Val, Leu, Ile, Pro, Phe, Tyr, Glu, Lys, and Cys and mean relative percent differences below 6.26%. Methionine was lower by the HFB isobutyl method in each case, as was aspartic acid. The arginine HFB isobutyl values were in excellent agreement for ribonuclease, lysozyme and β-lactoglobulin, but were lower for soybean meal and poultry feed.

A comparison of the total absolute amounts of amino acids found by the HFB isobutyl method and by IEC show extremely small relative differences for the five hydrolysates. These data closely parallel and support the packed column GC-IEC comparison described in Chapter 2, Volume I which concluded that GLC and IEC methods yield essentially identical results when applied to the same hydrolysate.

V. CONCLUSIONS

The success of GC for analysis of amino acids has largely resulted from two achievements; the development of straightforward, quantitative amino acid derivatization procedures and the development of chromatographic columns which separate and elute the amino acid

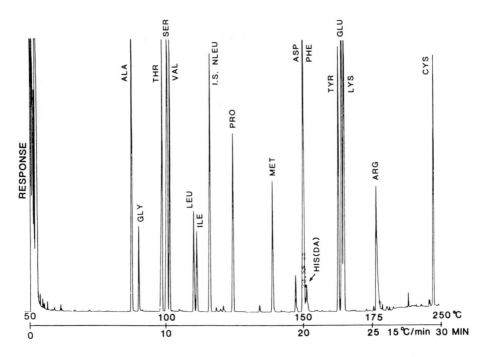

FIGURE 4.   Fused silica capillary GC analysis of amino acids *N(O)*-TFA *n*-butyl esters. Ribonuclease hydrolysate, final sample conc.: 400 μg/1.0 mℓ; injection: on-column, ~15 ng each amino acid (0.75 μℓ); Column: 30 m × 0.32 mm I.D., 0.25 μm film thickness; initial temp.: 50°C, 5°/min to 175°, DB-5 (phenylmethyl silicone) 15°/min to 250°C.

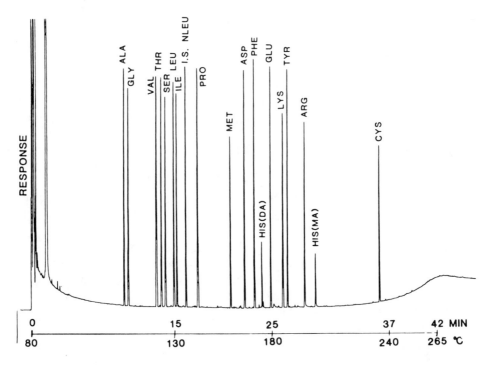

FIGURE 5.   Fused silica capillary GC analysis of amino acids *N(O)*-HFB isobutyl esters. Standard solution, final conc.: 20 μg each amino acid per 0.1 mℓ; Split injection, 40 per liter, 1 μℓ injected, each peak 5 ng; initial temp.: 80°C, Isotherm. 5 min, 5°C/min to 265°C; column: DB-1 (methyl silicone), 20 m, 0.25 mm I.D., 0.25 μm film thickness.

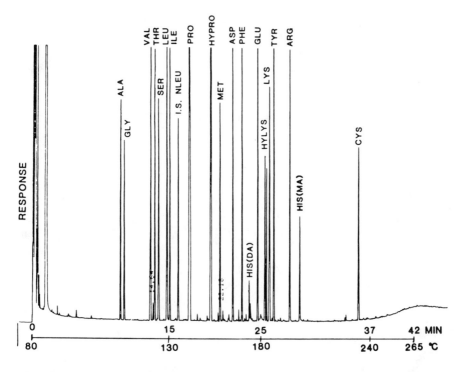

FIGURE 6.   Fused silica capillary GC analysis of amino acids *N(O)*-HFB isobutyl esters. Collagen standard, 250 nmol each amino acid per 0.1 mℓ, Pro, Hypro, 1250; split injection, 40/ℓ, 1 μℓ injected; initial temp.: 80°C, isotherm. 5 min, 5°C/min to 265°C; column: DB-1 (methyl silicone), 20 m, 0.25 mm I.D., 0.25 μm film ·hickness.

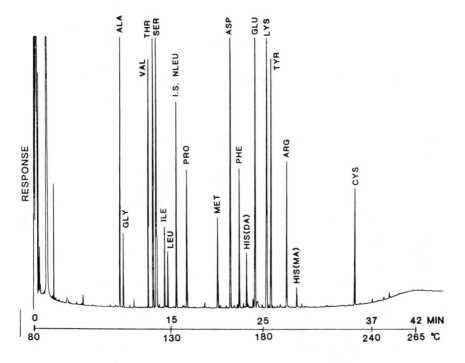

FIGURE 7.   Fused silica capillary GC analysis of amino acids *N(O)*-HFB isobutyl esters. Ribonuclease hydrolysate, final conc.: 20 μg I.S., 400 μg total amino acid per 0.1 mℓ; split injection, 40/ℓ, 1 μℓ injected; initial temp.: 80°C, isotherm. 5 min, 5°C/min to 265°C; Column: DB-1 (methyl silicone), 20 m, 0.25 mm I.D., 0.25 μm film thickness.

Table 5

COMPARISON OF FUSED SILICA
CAPILLARY GC AND IEC ANALYSES OF
FIVE DIFFERENT SAMPLE MATRICES[a,b]

	Mean relative % diff.[c]	
	TFA	HFB
Ala	2.69	2.60
Gly	2.87	2.06
Thr	1.69	1.66
Ser	2.70	3.00
Val	5.12	4.95
Leu	1.82	1.29
Ile	5.19	5.14
Pro	1.84	3.59
Met	10.63	10.78
Asp	NR[d]	9.61
Phe	NR	6.05
His	NR	32.40
Tyr	5.34	6.26
Glu	1.60	2.61
Lys	5.91	2.44
Arg	4.73	9.19
Cys	9.98	7.76
Total mean % diff.	4.44	6.55

[a] Ribonuclease, β-lactoglobulin, lysozyme, soybean meal, poultry feed.
[b] Samples hydrolyzed with 6 N HCl, 145°C-4hr, under vacuum; (see Experimental) three independent analyses of each hydrolysate were performed by each method. Each value is a result of an average of 15 results.
[c] Relative % differences are the GC values/IEC values × 100. Differences: TFA > IEC: 33, TFA < IEC: 33, TFA = IEC: 4; HFB > IEC: 42, HFB < IEC: 38, HFB = IEC: 5.
[d] Asp, Phe, and His peaks not resolved (NR).

derivatives. As presented in the previous chapters of these two volumes, a number of successful approaches have been developed and applied to a number of analytical problems. This chapter demonstrates the equivalence of capillary GC and IEC methods for determining amino acids in a broad range of different sample matrices. The *N(O)*-HFB isobutyl esters of the protein amino acids are easily separated and quantitated on a nonpolar methyl silicone column, whereas a column which completely separates the *N(O)*-TFA *n*-butyl esters continues to be sought.

The increasingly wide use of fused silica capillary columns with immobilized stationary phases for a wide range of analytical tasks now allows laboratories to perform amino acid analyses without the substantial investment of an expensive dedicated amino acid analyzer, an HPLC system specficially designed for amino acid analysis, or the preparation of column packings specific for amino acid analysis. With simply the purchase of a commercially available fused silica capillary column and the appropriate reagents, any laboratory with a capillary GC instrument capability can perform quantitative measurements of amino acids.

A column which provides the complete separation of the 20 *N(O)*-TFA *n*-butyl derivatives is not yet available, however, the *N(O)*-HFB isobutyl esters of the 20 protein amino acids are easily separated and quantitated on a nonpolar methyl silicone column.

ACKNOWLEDGMENTS

This research was supported in part by the Agricultural Experiment Station and the Missouri Research Assistance Act program.

REFERENCES

1. **Golay, M. J. E.,** in *Gas Chromatography 1958* (Amsterdam Symposium), Desty, D. H., Ed., Butterworths, London, 1958, 36.
2. **Lipsky, S. R.,** The fused silica capillary column for gas chromatography, *J. Chromatogr. Libr.* 32, (Sci. Chromatogr.) 257, 1985.
3. **Gil-Av, E., Feibush, B., and Charles-Sigler, R.,** Separation of amino acids enantiomers by gas liquid chromatography with an optically active liquid phase, *Tetrahedron Lett.,* 1009, 1966.
4. **Frank, H., Nicholson, G. J., and Bayer, E.,** Rapid gas chromatographic separation of amino acid enantiomers with a novel chiral stationary phase, *J. Chromatogr. Sci.,* 15, 174, 1977.
5. **König, W. A., Rahn, W., and Eyem, J.,** Gas chromatographic separation of diastereomeric amino acid derivatives on glass capillaries, *J. Chromatogr.,* 133, 141, 1977.
6. **Jellum, E., Bjornson, I., Nesbakken, R., Johansson, E., and Wold, S.,** Classification of human cancer cells by means of capillary gas chromatography and pattern recognition analysis, *J. Chromatogr.,* 217, 231, 1981.
7. **Schneider, K., Neupert, M., Spiteller, G., Henning, H. V., Matthaei, D., and Scheler, F.,** Gas chromatography of amino acids in urine and haemofiltrate, *J. Chromatogr. Biomed. Appl.,* 345, 19, 1985.
8. **Dandeneau, R. and Zerenner, E. H.,** Fused silica capillary columns for gas chromatography, *J. High Resolut. Chromatogr. Chromatogr. Commun.,* 2, 351, 1979.
9. **Grob, K., Grob, G., and Grob, K., Jr.,** Capillary columns with immobilized stationary phases. I. A new simple preparation procedure, *J. Chromatogr.,* 211, 243, 1981.
10. **Grob, K. and Grob, G.,** Capillary columns with immobilized stationary phases. II. Practical advantages and details of procedure, *J. Chromatogr.,* 213, 211, 1981.
11. **Labadarios, D., Moodie, I. M., and Shepard, G. S.,** Gas chromatographic analysis of amino acids in physiological fluids: a critique, *J. Chromatogr.,* 310, 223, 1984.
12. **Gehrke, C. W., Wall, L. L., Absheer, J. S., Sr., Kaiser, F. E., and Zumwalt, R. W.,** Focus: amino acid analysis. Sample preparation for chromatography of amino acids: acid hydrolysis of proteins, *J. Assoc. Off. Anal. Chem.,* 68, 811, 1985.

INDEX

A

B

C

F

G

H

I

W

Wall-coated open tubular (WCOT) column, 70—71,
    112, 138
Water sample, 77, 79

X

α-(2,5-Xylyl)ethylamine, 54